Stephen Hawking

Most Inspiring Scientist of the World

(Extraordinary Life Lessons That Will Change Your Life Forever)

Patrick Ellsworth

Published By **Darby Connor**

Patrick Ellsworth

All Rights Reserved

Stephen Hawking: Most Inspiring Scientist of the World (Extraordinary Life Lessons That Will Change Your Life Forever)

ISBN 978-1-9992226-7-3

No part of this guidebook shall be reproduced in any form without permission in writing from the publisher except in the case of brief quotations embodied in critical articles or reviews.

Legal & Disclaimer

The information contained in this book is not designed to replace or take the place of any form of medicine or professional medical advice. The information in this book has been provided for educational & entertainment purposes only.

The information contained in this book has been compiled from sources deemed reliable, and it is accurate to the best of the Author's knowledge; however, the Author cannot guarantee its accuracy and validity and cannot be held liable for any errors or omissions. Changes are periodically made to this book. You must consult your doctor or get professional medical advice before using any of the suggested remedies, techniques, or information in this book.

Upon using the information contained in this book, you agree to hold harmless the Author from and against any damages, costs, and expenses, including any legal fees potentially resulting from the application of any of the information provided by this guide. This disclaimer applies to any damages or injury caused by the use and application, whether directly or indirectly, of any advice or information presented, whether for breach of contract, tort, negligence, personal injury, criminal intent, or under any other cause of action.

You agree to accept all risks of using the information presented inside this book. You need to consult a professional medical practitioner in order to ensure you are both able and healthy enough to participate in this program.

Table Of Contents

Chapter 1: A Short History Of Stephen Hawking

Dennis William Sciama (1926 – 1999) end up a don on the University of Cambridge inside the United Kingdom. He became one of the most eminent physicists of his time.

In 1963 he become informed that he emerge as to get hold of a current pupil, a more youthful man from Oxford who desired to adopt his doctoral thesis beneath his tutelage.

There modified into now not something uncommon in this. Mentoring new college students emerge as element and parcel of a university educational's existence.

However the brand new pupil regarded, at the face of it, unremarkable. In fact he had the recognition of a lazy and truly difficult pupil.

In his written examination at Oxford he had performed neither a primary nor a 2d diploma. A first might have entitled him to undertake postgraduate research at Cambridge; a 2nd at Oxford.

He needed to located as a whole lot as an oral exam, a trial that terrified him but no matter the fact that stimulated his examiners who remarked that they confronted intelligence greater than there on.

After a while Sciama moreover agreed that he become dealing with a quite sturdy mind.

This guy have become handiest 21 years antique and furthermore had clearly been given 2 years to live.

His call modified into Stephen William Hawking.

Hawking modified into born on January 8 1942 in Oxford, to Frank and Isobel (nee Walker) Hawking. He has extra younger sisters, Phillipa and Mary, and an adopted brother, Edward.

Both dad and mom had been academics at Oxford University even as Stephen changed into born. Frank look at treatment while Isobel have a observe philosophy politics and economics.

Frank ought to emerge as head of the parasitology branch at the National Medical Institute, London, in 1950.

But on the time of Stephen's beginning the circle of relatives have come to be now not wealthy. Frank's grandfather had bankrupted the Hawkings by using incautious purchases throughout the Great Depression. It changed into best his companion that had saved them from utter disaster by using means of beginning a faculty in their domestic.

Isobel end up the son of a Glasgow physician who have become decided to ship her daughter to college at a time whilst an educational profession (or a profession of any type past being in company, nursing or education) changed into considered alternatively uncommon for girls.

Oxford University did now not even award stages to girls till 1920.

Perhaps it have become this scape with ruin that made Frank Hawking a specially meticulous and methodical more youthful guy.

Isobel modified into extra adventurous and hated being hemmed in.

When she have become in confinement at health center with Stephen (she changed into in her final week) she left the sanatorium and amused herself window buying inside the Oxford streets.

She went into a e book save and there presented an astronomical atlas. Reflecting at the profession her son would take, she regarded this as a propitious purchase.

The Hawkings had been a bookish and fairly eccentric own family. Often meals can be spent in whole silence at the equal time as Dad. Mom and Stephen have a take a look at at desk.

The family travelled in an vintage London cab, saved bees in the basement and made fireworks in the glasshouse. And Frank's uniqueness, tropical illnesses, should have made for captivating discussions over dinner, after they weren't all studying, of route.

Frank desired Stephen to comply with him into remedy, but Stephen had considered one of a type mind.

The boy cherished technological data, and changed into frequently positioned searching up on a starry night time, misplaced in surprise.

Stephen have become a very energetic lad. He have emerge as sociable, and loved to climb, play video games and to dance.

As a student he have become shiny, even though considered unexceptional. It may be that he considered the curriculum unchallenging, otherwise he did no longer have a strong sufficient attention for his thoughts. The latter would possibly appear like borne out through his college opinions.

The family moved to St Albans, a small city lying virtually to the north of London, while Stephen changed into eight.

He attended St Albans School. This all boys' institution is now price paying, and is one of the oldest unbiased colleges within the United Kingdom.

At the time of Stephen's education, but, it have grow to be an instantaneous provide college. This intended that a number of college students have been supplied locations on the premise of a success scholarship packages, at the same time as for others, regular charges have been paid.

St Albans at a few stage inside the 1950s became a mixed bag of a college. It sported the astonishingly beside the point motto 'Mediocria firma' – which translates as 'the center way is most steady'.

The motto was taken from the bi-sexual Elizabethan logician Francis 1st Viscount St. Albans, and brought on Stephen's headmaster – WT Marsh – agonies of internal contradiction.

The man changed into duty sure the exhort, the traditions of the dictum at the same time as being horrified with the aid of the usage of affiliation of the phrases with mediocrity – the remaining function with which a college together with St Albans could need to be associated.

(In reality, the adage related to the unstable Elizabethan times, whilst to stray from the middle ground may be seen as politically threatening, and may as like as not land the protagonist in essential, existence threatening, hazard.)

St Albans School in the Nineteen Fifties become a surely eccentric place. A combination of Masters and (now and again) Mistresses taught the men.

These instructors were intellectually extraordinarily in a position, and may properly have inspired the more youthful Stephen onto future academic brilliance, but additionally bizarre of their conduct.

The cane emerge as wielded considerably, and now not first-class through adults. Prefects have been perfectly at liberty to conquer boys who crossed barriers – or now not, in masses of instances.

In fact, an air of violence simmered beneath the surface of the vintage institution. WT Marsh ought to robotically lash out at university college students who indignant him – a left wing 6th form poet frequently bearing the brunt of his anger.

He as quickly as used the loss of lifestyles in a biking twist of destiny of an antique boy of the university as an opportunity for a short lesson inside the proper which means of tragedy.

He defined the more youthful man's loss of life as unhappy for his dad and mom, but now not tragic. That must were a fantastic consolation to them!

As a boy at St Albans, the younger Stephen have end up given to what in recent times we might bear in mind creative but slightly offbeat conduct.

Whilst prepared sports activities activities sports activities – as essential to faculties which encompass St Albans in recent times as they had been within the Fifties – held little attraction, he might create complex board video video games regarding delusion battles.

A type of forerunner to Dungeons and Dragons.

In a harsh surroundings, only the most effective live on. St Albans modified into such an area, as were many fee charging schools of the time.

Those who opted out of the army issue of training – it have become feasible to accomplish that, however hard – have been automatically humiliated. They have been given the assignment of building a Greek Ampitheatre at one point.

The assignment concerning now not excellent bodily very difficult work (undergo in thoughts, they were nonetheless pretty extra youthful boys) however to be completed dressed truly in shorts and t shirt – no matter the climate.

And notwithstanding the reality that every unmarried boy had exceeded a difficult the front exam to advantage get right of entry to to the faculty – even those no longer in receipt of direct investment – to be solid in the bottom go with the flow was taken into consideration educational failure.

Stephen became in reality not on this class. He sat truly 1/2 manner via the A devices. If this appears surprising for a man who would probable emerge as known as a genius, it's miles probably not that surprising.

St Albans and the encompassing region had visible a mass influx of European and Jewish immigrants for the duration of and after World War Two. Many of those households had knowledgeable their sons to gain success at a university together with St Albans.

They possessed likely a piece ethic and tolerance of the device that Stephen did no longer. The more youthful Hawking moreover had a totally eccentric thoughts, modern past the boundaries of the curriculum.

Whilst he absolutely engaged with the numerous very intellectual instructors, the curriculum of the time can also want to offer little notion to him, being geared nearly completely to winning an Oxbridge scholarship.

He changed into additionally a boy who favored to assignment the structures at St Albans. When older, he prepared ban the bomb marches to Aldermaston, wherein the UK's chemical and natural war guns inside the suggest time are evolved.

Whilst Stephen did adopt the officer cadet schooling at the school, he did so with a type of charming defiance, for instance via using refusing to maintain his uniform clever.

And while at the firing range, long term taking note of troubles have been far from the teachers' minds – it grow to be visible as a susceptible factor to require ear plugs.

However, quite glad to be appeared for this reason, Stephen created his very personal pair from blotting paper, which he crammed thus far down his ears that they had to be eliminated with the useful resource of the medical medical doctor day after today.

He as a small boy, and as such become a natural goal for the bullying that prevailed on the college.

It appears as even though he took it properly sufficient, counting on his quick wit to vicinity down the bigger boys, and accepting that once in a while he would probable escape, once in a while not.

In this simmering cauldron of a college, in which explosions have been continuously just for the duration of the corner, Stephen drew idea from his math grasp, Dick Tahta.

With the Armenian teacher, he constructed an early laptop from discarded electronics.

And it's far hard to argue toward the environment of the school main Stephen on his adventure to turning into the genius that lives today.

Although, it'd have been his choice to be precise from the norm that supplied the greatest thought.

His dad and mom were keen to provide him an training as that that they had had, and he enrolled in the University of Oxford in1959, on the equal time as he modified into quality 17.

Oxford University is the oldest college in the English talking, relationship decrease returned to at least 1096 but likely an entire lot older than that.

The university turned into so prestigious and critical in the life of the UK that until 1950 it elected members to the British House of Commons.

The University has 38 faculties, to which within the Oxford and Cambridge Universities college students ought to be affiliated.

Hawking joined the oldest of the Oxford colleges, University College. That college produced such women and men because the British Prime Minister Clement Atlee, US President Bill Clinton, the poet Percy Bysshe Shelley and the violinist Sophie Solomon.

Hawking cherished college existence. He mainly cherished the social lifestyles, as of path many college college college students do. He joined a rowing crew as a coxswain. It turn out to be said that he turn out to be a instead enthusiastic, if no longer reckless, coxswain, dropping oars and getting the organization into risky scrapes.

And as is the case for lots college university college students his research tended to take 2d location.

He grow to be brilliant and he knew it. He want to jot down essays and take assessments with a minimal try and regularly with little, if any education.

While this got him through college he suffered from lack of software application. His professors and tutors would comment on his lack of recognition and carelessness.

It have become best in his final yr, 1963 that he found out he needed to cognizance on his energies and make up his mind what he wanted to do with an Oxford schooling.

Chapter 2: Reaching For The Stars

Sir Fred Hoyle (1915 – 2001) grow to be a Cambridge University don and one of the essential, if now not the eminent astronomer of his day.

At a time while cosmology, the look at of the origins and evolution of the universe, have become in its infancy Hoyle emerge as propelling it ahead.

The concept that the universe turned into tens of hundreds of thousands of years vintage and superior through physical prison suggestions and no longer thru way of the direct intervention of a deity modified into but quite new.

In 1897 the famous physicist Lord Kelvin had proposed that Earth itself changed into probably spherical 20 million years vintage. We now understand that estimate is ludicrously short, not nearly prolonged sufficient to account for the formation of the arena and evolution of existence. Today scientists take delivery of as true with that close to 4 billion years is a more accurate estimation.

But how vintage have come to be the universe itself?

Hoyle went issue manner to calculate an age via way of his artwork on stellar nucleosynthesis, this is, the formation and evolution of stars.

He confirmed that stars have been shaped through the coalescence of fuel clouds under gravity. These dense clouds produced a big nuclear reaction that normal a film big name.

Moreover, as those stars run out of gas they enlarge into such entities as pink giants, and then crumble into smaller stars with big gravitational pull, which embody white dwarves.

This evolution established that the universe needed to be many billions of years antique. Our Sun, for instance, long-established approximately four and a half of of billon years inside the past and is anticipated to last enough 10 billion years.

But stellar nucleosynthesis does not in itself inform us how antique the universe itself is, nor does it solution an apparent accompanying question – did the universe have an beginning region and if so, how did it come into being?

This is of direction the question that has absorbed humanity from its very inception. For centuries it belonged to the geographical regions of faith and philosophy.

With the arrival of technological facts as we apprehend it in 16th and seventeenth century Europe were able to start to solution how the universe functioned however notwithstanding the truth that now not the manner it got here into being – if in any respect.

Its life have become though attributed to some supernatural modern precept, whether or not or not or no longer it have emerge as a personal God or an impersonal character who created and sustained the universe however with out taking any hobby in it.

In 1917 Albert Einstein proposed that the universe modified into static and basically unchanging, implying that it had no basis.

Later he modified his mind after the well-known astronomer Edwin Hubble (1889 – 1953), after whom of path the gap telescope is called, positioned that apparent clouds of fuel in area referred to as nebulae have been in reality galaxies, and seemed to be receding into space.

From this he deduced that items in the observable universe have been sincerely expanding away from every one-of-a-kind.

Einstein adjusted his private precept on the way to account for this, however nevertheless the apparent increase couldn't be satisfactorily described.

Enter Fred Hoyle. He agreed with Einstein that the universe end up in reality basically static and unchanging.

Stars had been born and died, of course. But as a body is ordered and balanced through physical prison hints; the universe remained static. He known as this model of the universe the Steady State Theory.

How then did Hoyle deliver an cause for the plain boom of the universe? He intended, with none definitive proof, that consider modified into usually being created in the spaces among galaxies.

So the universe modified into some thing like a pool of paint expanding because of the fact paint from a can is being continuously and calmly poured into it.

The Steady State Theory neatly avoids the hassle of the starting vicinity of the universe with the aid of mentioning it never had one. The universe is eternal.

Apart from exceptional scientific objections that are in all likelihood too complex to discover now, the concept does now not solution the obvious query of strategies can some factor be constituted of not whatever?

This of direction is a query that theologians are requested as well. And it changed proper right into a question that have turn out to be to be – and stays – the point of interest of Hawking's studies.

Nevertheless Steady State Theory turned into drastically time-commemorated at some point of the 50s and 60s, the use of in large component on the popularity of Hoyle.

Understandably Hawking have become disturbing to worrying to stable Hoyle's tutelage for his doctorate thesis.

However it changed into not to be.

Besides being a large of astronomy and physics Hoyle have become some factor of a scientific movie big name, lots as Hawking would be.

He became continuously journeying and giving lectures, being interviewed, writing, and using his recognition to interfere in fields often outdoor of his degree of information.

For instance, he famously entered the lists closer to paleontologists through in 1986 claiming that fossils of Archaeopteryx, a creature with affinities of each birds and reptiles and for that reason appeared as transitional a few of the 2, were faux.

In the Seventies Hoyle proposed that existence did now not originate on Earth however as an opportunity developed from microbes in meteors and comets which have struck Earth.

He apparently suggested that unexpected and violent pandemics which incorporates the 1918 Influenza Epidemic and the 1986 Mad Cow Disease outbreak in the United Kingdom were proof of alien assaults.

Hoyle might also need to have made an mistaken coach, being tough to pin down and susceptible to occurring tangents and fanciful ventures, none of which undermined his undoubted medical prowess within the place of physics.

And there also can had been every other cause why Hoyle did now not take younger Hawking underneath his wing.

On one occasion Hawking modified into taking note of Fred Hoyle lecturing on the Royal Society. At one aspect he interrupted Hoyle to accurate him on some thing he pronouncing.

Hoyle turned into astonished to pay attention a person who had barely graduated hopefully formidable to correct him, and have to had been even more astonished to discover that Hawking was proper.

So extra youthful Hawking were given Denis Sciama.

Sciama had precise advantages over Hoyle that Hawking definitely preferred at that time. Sciama end up exceptional, like Hoyle, however become now not a celebrity. He became reliable.

Further he end up absolutely solicitous for his college students. He emerge as encouraging and supportive, and diagnosed Hawking's precise intellect and moreover noticed that it had to be advanced.

Hawking modified into in sore need of encouragement, for he had just been recognized with Amyotrophic lateral sclerosis (ALS), additionally called Motor neuron illness or Lou Gehrig's contamination.

This discomfort is characterized via the typically fast deterioration of muscle feature predominant to loss of life.

The motive of the sickness is unknown and there may be no remedy.

Hawking end up already beginning to lose manipulate of his frame. He had falls and his speech have become starting to slur.

With his medical docs giving him 2 years to stay and faced with a horrible and humiliating decline, Hawking have become depressed and in want of a purpose to hold on.

He placed comfort in his wife, Jane Wilde. The had met at a News Years birthday party at Cambridge. They have turn out to be engaged in October 1964 and married on July 14 1965.

Hawking said that the wedding gave him 'some thing to live for', but in advance than this Sciama encouraged the depressed more youthful guy to reputation on his thesis.

Hawking selected as the topic for his thesis a idea of the begin of the universe – an ambitious task for a graduate!

He decided on to contradict the mind that he had preferred as his train – Fred Hoyle. He did now not preference to protect Steady State Theory. Rather he decided on to signify the opposite amazing concept of the inspiration of the universe – the Big Bang.

The father of the Big Bang idea became a Belgian priest, physicist and mathematician, Georges Lemaitres.

In 1927 Lemaitres described the apparent increase of the universe with the useful resource of positing that each one depend had first off been condensed in a unmarried issue – a singularity.

This singularity then improved, forming the universe which stays dashing outwards from the aspect even as the singularity 'exploded'.

This concept got here to be determined through some of physicists and astronomers who believed it awesome organized the observable records.

The principle had the gain of now not having to provide an reason behind the advent of new depend, for all of the rely that ever turn out to be contained in the primeval singularity.

However it couldn't deliver an purpose for what brought approximately the preliminary increase of the singularity. If all depend grow to be contained in the singularity then what brought on it to blow up?

Further it did not offer an reason of in which that matter amount comes from in the first region.

We casually and not virtually efficiently use the word 'explode' however ironically Fred Hoyle used this terminology to sentence the idea.

In a 1949 radio interview he mentioned the idea of a universe increasing outward from a unmarried component because the 'Big Bang;' concept. And of direction the decision has caught, the concept fiercest opponent virtually giving the precept its call.

A critical problem for the Big Bang theorists emerge as supplying a mechanism. How should the complete mass of the universe be contained in a unmarried issue? There modified into no version, and no observable item within the universe that could provide a clue.

Nevertheless Albert Einstein had envisaged the possibility of singularities in his 1915 General Theory of Relativity.

In massive phrases the General Theory of Relativity holds that there can be no regular object in the universe. Everything is in movement and is in motion relative to the movement of the whole thing else. So a chicken flying actions in phrases of the movement of the air, which in flip is relative to the motion of the earth, the movement of the sun, of the solar device, the Milky Way, and so forth.

From this pretty simple remark Einstein extrapolated requirements for time and rely.

For example, he concluded that as there has been no constant object, there has been no regular aspect of time every. Time turned into fluid.

From his General Theory of Relativity Einstein got here up with a Special Theory of Relativity, which stated that power and mass have been collectively convertible. He composed the famous equation $E=MC^2$; an item is convertible to an quantity of electricity same to the mass of that item accelerated thru the usage of the charge of moderate squared.

We see an lousy demonstration of this principle in nuclear fission, at the same time as an atomic bomb explodes, idea General and Special Relativity have a plethora of applications in the situation of physics.

The young Hawking set his thoughts to growing use of those mind to the idea of a singularity. His reason have become grandiose. He desired to illustrate that singularities want to exist and in reality do exist, and that the start of the entire universe from this kind of singularity may be satisfactorily defined.

In this he changed into stimulated with the useful resource of the art work of Roger Penrose (b. 1931), the English mathematician, physicist and era philosopher.

Penrose had written a paper postulating that the gravitational pull on the middle of a black hole might be so insuperably sturdy that rely, location and even time itself was compressed right proper into a singularity.

Now for us black holes are the common stuff of era fiction films and novels, however while Hawking have come to be writing his thesis they have been little greater than mathematical curiosities that have been postulated by using manner of the General Theory of Relativity.

Indeed the term 'black hole' modified into only accompanied as a handy descriptive in 1964. At the time there has been a lot theorizing about what the person of a black hole is probably.

Physicists can now tell us how weird black holes are, regardless of the truth that they are despite the fact that shrouded in mystery and there can be nevertheless a good deal extra they don't recognize about them than what they do apprehend.

A black hole is a collapsed superstar, even though not all stars become black holes. They rise up from huge stars like crimson supergiant, which explode as a supernova and then regularly disintegrate into super-dense hundreds that exert so effective a gravitational region that no longer even light can get away.

Strange subjects arise in black holes. The regular felony pointers of location and time appear to disappear.

Einstein's General Theory of Relativity predicts that gravity curves depend. So if a person were to fall right proper into a black hole they'll seem like stretched out like a piece of elastic.

Now gravity no longer most effective curves place, but time too. So a few factor weird might also take location to time too.

The nearer an item movements in the route of a black hole the slower it travels in time. When that object reaches the occasion horizon, the boundary beyond which even mild can not get away, time would possibly slow loads that it'd seem to freeze – all of the time.

However, all that is what an observer might see. They might see, at the component of the whole lot that modified into ever drawn with the aid of the use of the black hollow, someone stretched out and frozen in time.

On another degree, the individual couldn't exist for they could had been vaporized by way of way of manner of the awful power of the black hollow.

But count on that individual ought to stay on, and plummeted into the depths of the black hollow?

No-one is aware of. At the middle of a black hole region and may be infinitely curved, turning into Penrose's singularity. The felony hints of the universe as we understand them couldn't exist, and nobody is aware about what that might be like.

In his thesis Hawking proposed that the hypothetical singularity in a yet hypothetical black hollow may want to provide the precise version for the Big Bang.

Suppose that every one the trouble that is or ever changed into within the universe modified into compressed in a unmarried aspect. Then that singularity exploded, releasing the trouble and shaped the universe.

For us the concept of the Big Bang is firmly embedded within the well-known creativeness, even though we have to remind ourselves that at the time the Big Bang come to be debatable, and the life of black holes modified into not installation till one turned into observed in 1971.

Hawking's thesis have emerge as consequently formidable and scary.

It turned into so inspiring in truth that once acquiring his doctorate Hawking and Penrose decided to art work similarly almost about singularities and the opportunity of the universe originating from one.

In 1970 they collectively published a evidence that the universe have to have derived from a singularity.

Further theories came thick and fast. With physicists James Bardeen and Brandon Carter he composed a fixed of laws governing black holes.

Hawking defined his studies on black holes in his first e-book, entitled The Large Scale Structure of Space-Time, which he wrote with the mathematician and cosmologist George Ellis in 1973.

In 1974 Hawking carried out what have become perhaps his finest fulfillment inside the clinical place. In a paper he asserted that the proposition that now not something may also need to escape the gravitational pull of a black hole is not virtually actual.

According to Hawking, black holes emit radiation, very slowly and almost imperceptibly. After many eons the lack of power reasons them to scale back. As they cut lower back they grow to be hotter.

Eventually they become so warm that they explode, and all of the depend that has been subsumed into the black holes is released again into the universe.

At first this concept have come to be controversial as it contradicted the extensively significant concept that a black hollow modified into an absolute finality.

The concept of Hawking Radiation, as his idea is known as, sees black holes now not due to the fact the ultimate destroyers inside the universe however as engines of advent. Black holes offer upward push to new stars and galaxies. They are, if you could, the terrific recyclers of the universe.

Hawking Radiation additionally solutions a query about the universe's probably origin from a single point of singularity, just like a black hole.

If the singularity were a aspect of infinitely dense rely from which no longer satisfactory mild need to escape, how want to the initial explosion possibly upward push up? What stress need to triumph over the gravitational pull?

Hawking Radiation smartly avoids that hassle by using manner of pointing out a black hollow loses definitely charged atomic debris on the event horizon, the aspect round a black hollow past which the gravitational pull of a black hole is overpowering.

Negatively charged atomic particles keep into the black hollow singularity.

The concept has become widely ordinary, and it earned Hawking a Fellowship within the famous Royal Society of teachers and scientists.

As but there is alas no direct proof of Hawking Radiation. That is why Hawking has not as however been supplied the Nobel Prize for this concept.

Following his speculation about Hawking Radiation Hawking speculated approximately a horrible opportunity. He stated the opportunity that a collapsed black hollow also can very well wipe out the entire universe.

In a 1993 studies paper Hawking supposed that now and again a black hollow that without a doubt evaporates would leave a singularity. A singularity is a factor of limitless density.

This singularity is probably 'bare', meaning that the event horizon – the thing beyond which not even mild can get away – has disappeared moreover.

This approach that the awful electricity of the singularity is uncovered. There isn't always anything to prevent the entire universe from being ripped apart via using the abruptly spreading singularity

Now black holes take many billions of years to evaporate, and possibly none ever has. On the opportunity hand, likely a bare singularity is presently hurtling within the course folks.

Even so, it'd take many tens of thousands and thousands of mild years to reap us, so we can possibly but ebook that holiday.

Hawking has stated he prefers the speculative technique in theoretical physics. He prefers to postulate an idea and notice if the records and the arithmetic in form his idea.

This way that sometimes he has been incorrect, and on a few quite essential subjects. But he has normally been prepared to re-compare the statistics at his disposal and admit his errors.

Hawking seemingly has a penchant for making wagers along with his colleagues over scientific questions.

For instance, he guess a colleague, Kip Thorne, in 1975 that Cygnus X-1, a supply of X-rays in the Cygnus constellation, became no longer in fact a black hollow.

By 1990 it had emerge as broadly usual, based totally totally on statistics that it changed into a black hollow and Hawking conceded the wager.

Thorne's prize emerge as a 365 days's subscription to Penthouse magazine.

In 1975 Hawking became appointed Reader in Gravitational Physics at Cambridge University. By this time his speech have turn out to be swiftly deteriorating, and he been the use of a wheelchair for several years.

He refused to concede a few component to the sickness that ravaged his frame, and his family regularly remarked that he appeared to pretend that it did not exist, every now and then to their frustration.

He become infamous within the wheelchair, recklessly careering down the passages of Cambridge and once in a while on foot over feet.

After contracting lifestyles-threatening pneumonia in 1985 a tracheotomy emerge as done on him and he out of place what remained of his speech.

This necessitated round-the-clock care and the employment of nurses jogging in three shifts.

It also intended that opportunity techniques of conversation needed to be devised.

At first gambling cards had been used, but in 1986 his well-known voice synthesizer became devised and hooked up.

The synthesizer has an American accessory as it became made in the United States. Hawking regards it as his personal voice, and refuses to apply a few other. In truth, he has patented the voice so no-one else may additionally use it.

At first he selected letters thru way of shifting a hand. However he has for the reason that out of location motion in all but his cheeks and he uses his proper cheek to manipulate the device.

From the severa, many interviews we see recorded with Stephen Hawking, it seems as even though his speech device works similar to a human voice.

Of route, at the same time as one considers the hassle, this may now not be proper.

In fact, speech can be very gradual, as he controls the device with facial muscle businesses to carry out the sounds.

The affect we see within the course of interviews comes approximately due to the fact Stephen has the questions earlier, and is able to 'pre-document' his solutions to them.

Though his body declined his mind remained sharper than ever, and he regarded able to inhabit a worldwide of pure thoughts. Often he may spend hours through using himself. Afterwards he ought to say 'I even have

Chapter 3: Prisoner

In the 80s Stephen and Jane's marriage grow to be breaking down.

Academically they were of different minds. Stephen thrived inside the worldwide of mathematical readability and perfection. Jane cherished the comparative freedom and creativity of the arts. She cherished track as properly. Neither in reality understood the pleasures the others located in their disciplines.

Jane had a doctorate in Spanish poetry, and had felt compelled to collect it truly so she may probable have an academic lifestyles at Cambridge super from that of her husband.

Jane modified into deeply religious. Stephen end up, and remains, a business enterprise atheist. At times he mocked her faith.

Jane discovered his devotion to the 'goddess Physics' ironic, for the purpose that it modified into she who made the selection to maintain her husband on lifestyles useful resource even as he reduced in size pneumonia.

The scientific medical doctors had given him up for useless, but Jane had religion that he want to stay on.

Jane formed a strong however platonic relationship with an organist from the church choir she belonged to. His call emerge as Johnathan Hellyer Jones. Jae might later marry him.

This relationship modified into apparently ordinary through Stephen.

Though committed to her husband, Jane located getting to the goals of Stephen wearying, mainly as he regarded oblivious to his condition and to the goals it made on others.

At period he did agree to lease greater help. University college college students took on the characteristic of personal assistants and nurses undertook his physical care.

Though this did loose Jane to pursue her personal interests it also created a modern strain on their dating.

Accompanied through aides for max of his waking hours Hawking become developing remoted from his spouse's effect.

As carers often do, Hawking's assistants superior a sense of obligation that resented and excluded the impact of others, even that of his spouse and children.

One carer in particular turn out to be assuming the vicinity due to the reality the number one mother or father of Hawking and interpreter of his wishes.

This individual modified into Elaine Mason. She joined Hawking's nursing crew within the 1980s. She unexpectedly assumed the primary function inside the stressful group, other than all however her very private have an effect on.

Hawking and Mason have end up close to, and in 1995 he married her.

After their marriage Hawking's kids and pals observed themselves unable to method him with out going via Mason, who regarded to manipulate each issue of Hawking's life.

Then, in 2000, Hawking supplied to a network health center with a broken arm and a split lip. He refused to offer an reason behind how he had come by the use of them.

In January 2001, truly shy of his sixtieth birthday, he offered over again with a broken femur, pronouncing that he had crashed into a wall.

To people who knew him this seemed great, given that he modified into so adept in his wheelchair.

Many in Hawking's circle, which consist of a number of his nurses, suspected Mason. They pointed out taking note of her scream abuse at her husband.

The police were known as in via his daughter, Lucy, however Hawking refused to reveal how he sustained a string of injuries, decrying the intrusion into his private existence.

On March 29 2004 Cambridge police dropped their research of Hawking's injuries, putting ahead it to be 'extraordinarily thorough.'

However it have become referred to that handiest 12 people were interviewed, and that Hawking himself vehemently denied the allegations that he had been physically abused.

'I firmly and wholeheartedly reject the allegations that I had been assaulted,' he said. In the moderate of those denials the investigators decided themselves stymied.

Friends of Hawking had been astonished. One worried man or woman had recommended a communique with him approximately Elaine. 'I can not be left on my own together with her,' he said. 'Please don't skip.'

Why then did Hawking refuse to confess to even his youngsters that he turn out to be being assaulted? Given his attitude closer to his debility, it could be that he did not need to attract hobby to his vulnerability.

Perhaps the hold Elaine Hawking had over him grow to be too sturdy, and he feared dropping her. Relationships of dependency some of the abused and their abusers are not unusual.

On October 20 2006 Stephen and Elaine agreed to divorce. Hawking refused to comment on the divorce, concerning media interest and unwanted distraction, and to in recent times he has now not spoken about the accidents he sustained in the course in their marriage. There isn't any new studies.

After their separation Hawking drew inside the direction of his three kids and truely to Jane, who thru now turn out to be married to Hellyer Jones. Jane always remained interested by and concerned for her ex-husband, and that they stay close buddies.

A Theory for Everything

Even in the path of his personal trials Hawking remained targeted on his instructional pursuits. By his sixtieth birthday he had his mind set on the most bold intellectual cause of all – a precept of the whole thing.

The Theory of Everything is the holy grail of theoretical physics. It is conceived as a form of all-encompassing framework that would deliver an cause of how the whole lot within the universe works and suits collectively.

So a ways scientists have components of a evidence. The General Theory of Relativity is one detail. Quantum mechanics, the observe of subatomic particles and their interest, is a few one of a kind. The four legal recommendations of thermodynamics governing electricity are any other.

The ultimate explanation, the glue that binds the myriad of physical criminal tips together, is but lacking.

Theoretical physicists aspire to offer you with a mathematical equation as a way to answer everything. It may also need to supply an reason behind how the universe got here into being. In particular it might provide an explanation for in which the initial depend inside the got here from. It could offer an motive at the back of how something got here from not anything!

To many, discovering this equation would be the pinnacle achievement of the human race. Scientific data can be complete.

Others are demanding approximately the search. They say that accomplishing such know-how might be like eating the fruit from the Tree of the Knowledge of Good and Evil inside the biblical Garden of Eden. We would probable come to be like gods, however horrible and immature within the manner we exercise divine power.

For Hawking but there may be no longer some thing to worry. The quest for the Theory of Everything is for him the destiny of the human mind, which on Earth is privileged to mirror upon the wonders of creation and ponder their reason.

He declared that a principle of the whole lot, or Grand Unified Theory, as it's also called, was on the horizon in Eighties. He said that it had a incredible danger of turning into a reality through the start of the 21st century.

The e-book of A Brief History of Time in 1988 marked Hawking's first assignment into the geographical areas of famous technological expertise.

He come to be approached by using manner of the usage of Cambridge University Press to write down a e-book for laypeople approximately the universe and its evolution, using simplified terms and explanations.

Hawking have become within the starting reluctant however he desired the coins. He have become first of all pissed off through the editors, who insisted he preserve simplifying his language.

The e-book popularized the paintings of Hawking, who till now had been a touch appeared scientist past the halls of academia.

Hawking rapidly rose within the estimation of his buddies and in 1979 he grow to be appointed Professor to the Lucasian Chair of Mathematics at Cambridge.

This august seat of learning became based via one Reverend Henry Lucas in 1663, and has been occupied via using such highbrow giants as Isaac Newton and the 19th century inventor of the computer, Charles Babbage.

Hawking's research style is, by his personal admission, greater intuitive and speculative than evidence – based totally honestly.

In one of a kind terms, he makes informed speculations that appear to suit modern-day statistics and appears for the proof to verify the idea, in desire to draw strict mathematical conclusions from the winning information.

'I could rather be proper than rigorous,' he has stated. For a scientist this could seem a as an alternative informal manner to are searching for for fact. Hawking but sees this as progressive. Many of his colleagues had been stimulated by his thoughts and scour the sky looking for evidence.

He has always been short to widely recognized while the proof has proved him incorrect.

One example of Hawking strolling into controversy with this method involved a problem known as the black hole statistics paradox.

All matter within the universe has information. By statistics scientists mean the mathematically quantifiable trends that depend possesses − period, mass, width, temperature, and so forth.

When depend reaches the occasion horizon of a black hole it's miles vaporized, as we have visible. Yet the information of the hassle remains.

We will keep in mind while we had been discussing black holes. The closer an item, say a spacecraft, techniques the occasion horizon the slower time will become. So at the same time as it movements the horizon an observer would possibly however see the spacecraft for all time, even though it is really destroyed.

In exclusive terms, the data quantifying that count might stay.

It's a tough concept to get the thoughts round. Even scientists have problem explaining it. But it makes revel in mathematically given Einstein's theories of relativity.

However Hawking proposed in his precept of Hawking Radiation that a black hole very step by step evaporates, and that subsequently it will disappear absolutely.

But if that passed off, in which does the records of the spacecraft skip? Does it disappear as well?

This became a hassle for physicists, due to the reality it's miles a nicely-based totally dogma of physics that statistics cannot be destroyed.

Hawking broadly diagnosed the anomaly and got here up with an answer. He recommended that the statistics contained in rely falling right proper right into a black hole modified into preserved and launched lower lower back into the universe thru the outgoing Hawking Radiation.

However, this is a hunch on Hawking's element. It is unproven, simply as Hawking Radiation itself stays unproven. Mathematically it's far fun idea for masses physicists. Yet there remains no concrete proof to verify the idea.

Hawking additionally courted controversy with the physicist Peter Higgs, over the latter's postulation of a ordinary area which gives the whole lot within the universe mass.

Hawking strongly disputed the life of the kind of area, until a particle of the area, the Higgs boson, changed into placed with the aid of the powerful Hadron collider in Switzerland in 2013.

The so-known as 'God particle' gives being to everything I the universe. Peter Higgs become presented the Nobel Prize for predicting the lifestyles of the Higgs boson and Higgs subject in 2013.

Stephen Hawking grow to be without a doubt satisfied that this shape of particle could not exist, or even guess with a colleague, Gordon Kane, that it might in no manner be decided.

Even after it grow to be discovered he lamented its lifestyles, complaining that it made physics a good buy much less interesting.

He even advised that the Higgs particle may want to recommend the literal doom of the universe. While admitting the existence of the Higgs Field he suggested that it'd in the end end up volatile ad vaporize the universe.

While this will sound like sour grapes on Hawking's element, he has truely in no manner been so petty as to disclaim proof-primarily based totally discoveries with the resource of his colleagues, even though Peter Higgs changed into supplied the Nobel Prize.

The discovery of the Higgs Field does but highlight Hawking's academic style and give an explanation for why he has never acquired the Nobel Prize himself, and probable never will benefit that honor.

Hawking's studies is mainly creative. He prefers to invest. Nobel Prizes are excellent given for proven medical discoveries. Hawking Radiation, Hawking's satisfactory speculation, even though substantially everyday by using the use of scientists, stays an albeit quite knowledgeable mathematical hypothesis. No proof has ever been observed to verify or deny its existence.

Hawking's innovative thoughts impels him to contribute beyond his area of theoretical physics. This is ironic, for the reason that the scientists who could not be his postgraduate mentor, Fred Hoyle, proved wrong precisely because of the fact he too ventured into extraordinary geographical regions.

Like Hawking, Hoyle modified into often earlier than a digital camera offering era to the public.

Hawking has warned humanity approximately developing an synthetic intelligence that is probably self-conscious. Such an entity, he says, may count on an lifestyles separated from humanity, evolve via itself and in all likelihood enslave or damage humanity in a horrendous Matrix or Terminator scenario.

He speaks about climate exchange and overpopulation and the diminution of Earth's belongings.

He believes that Earth' belongings is probably fed on within spherical a hundred years and then the human species will face extinction.

His solution is to colonize each different international, and so he encourages the exploration of different sun systems and the improvement of technology to move people to habitable planets in them.

The concept isn't always so outlandish as it may seem within the beginning. Recent astronomical observations have indicated the life of planets orbiting close by stars. Some of those planets are much like Earth in that they may be now not too close to their solar and no longer too a ways away as to make lifestyles not possible.

Hawking is a eager supporter of the Breakthrough Initiatives, a chain of applications based totally advert funded thru the Russian entrepreneur Yuri Milner I 2015 to are trying to find out liveable worlds and increase the era to colonize them.

The Initiatives moreover choice to discover extraterrestrial lifestyles, and Hawking appears a brilliant deal lots much less eager on this assignment.

Chapter 4: Instead He Claims That There Are 5 Reasons

His thoughts are based on the supposition of minute, unobservable subatomic particles referred to as strings. These strings vibrate in considered one of a type modulations and each modulation represents a terrific kingdom of being. And so the equal string should have some of trade and simultaneous realities.

Each string idea has eleven dimensions, this means that that there are multiple change realities inside the universe, every as true because the opportunity, and consequently a couple of reasons of fact.

This notable and nearly incomprehensible concept has been referred to as M-Theory. The call is enigmatic, and the M has been interpreted as reputation for 'magic' or 'thriller.'

However, like every Hawking's splendid mind, it stays an knowledgeable hunch. It excites physicists, plenty of whom are striving to confirm it and treatment a number of the troubles it offers.

From the ones thoughts Hawking extrapolated one approximately the beginning of the universe. It is useless, he postulated, to attempt to determine the muse of the universe, as it had a mess of numerous origins.

This is based mostly on the unusual technological expertise of quantum mechanics, the workings of debris smaller than atoms. There the prison suggestions of physics as we realize them in the macro international do not study. There is an entire new set of prison tips that physicists are however simplest starting to find out.

In the mysterious international of quantum particles exquisite realities can co-exist. Indeed, this new discipline of research is exploding antique thoughts approximately the character of place and time.

It has been advised that point itself isn't linear, flowing from the past through the winning and into the future like a circulate. Rather it can be the beyond and destiny happens element with the resource of aspect in trade however co-modern realities.

Time journey is every distinct famous issue on which Hawking has been satisfied to talk.

He says that adventure into the future is definitely possible. Indeed, it takes vicinity all the time.

The General Theory of Relativity holds that the nearer an object strategies the rate of moderate the more it moves into the destiny relative to its place to begin. In different phrases, time for that object s slows down.

So if a rocket or spaceship can be powered through a strength that would approximate the speed of light (regular with the concepts of relativity an object cannot supersede the speed of mild) it could theoretically assignment an occupant in advance in time.

Yet there will be a trouble, in that the faster the craft went, the extra mass it would tackle .If a craft were theoretically able to fly at just short of the rate of mild it might be so massive that the universe ought to scarcely incorporate it.

What of visiting within the past? According to the Einstein's General Law of Relativity, time can only tour forwards, now not backwards.

However, there can also theoretically be a way to adventure via time. Hawking and exclusive theoretical physicists speculate that a black hole may also want to create a wormhole.

A wormhole is probably a hyperlink amongst elements in space-time. But scientists have no concept how you will use this shape of portal. Firstly you may need to in a few way avoid being overwhelmed via way of the sizeable strength of a black hole.

Secondly, there may be no way to expect what may take area, nor how a wormhole may be engineered to move someone or item to a given factor in area and time.

And notwithstanding the reality that it have grow to be possible to harness the electricity of a black hollow to adventure again in time may want to we want to?

Travelling backwards in time may additionally need to create paradoxes. Suppose someone goes again in time and inadvertently kills his or very very personal father in advance than he end up born?

If he kills his father he can not exist. How then should he have lengthy long past lower back in time and killed him?

Hawking does not like that possibility. It disturbs the idea of an ordered, logical universe, and he has invented a law to save you it occurring.

Tongue in cheek, he has proposed that quantum debris may want to right now incinerate a time system to save you this form of paradox being created.

But in everyday style he offers a simple fact to push aside the concept of journey into the past. If it's miles feasible to excursion backwards in time, wherein might be the time vacationers? Where are the vacationers who've come to visit the past?

Perhaps they've got come, and we've were given had them sectioned. Or likely they're in cowl, below strict instructions not to intervene in data.

Hawking completed an check to decide as quickly as and for all whether or not or not time journey into the past have become feasible.

He held a celebration for time travelers. He posted the open invitation on YouTube. It have a take a look at 'You are cordially invited to a reception for time vacationers hosted by using way of Stephen Hawking to be hung on the University of Cambridge.'

He protected geographical co-ordinates and the date, adding 'no RSVP required.'

However, he did now not post the invitation till after the date.

He organized a room with balloons, champagne and nibbles and waited.

Alas, no-one grew to turn out to be up.

World's Greatest Living Genius?

In the famous imagination Stephen Hawking is probably the finest residing thoughts. Indeed one might be tough-pressed to don't forget every other scientist who has captured the eye of the arena as masses as Stephen Hawking.

There is the physicist Brian Cox, well-known as a presenter and writer of popular technological expertise. However his very own contribution to physics is noticeably small.

Likewise if we bypass past physics and go through in mind biology, we are able to hold in thoughts Richard Dawkins, regarded typically no longer for generation the least bit however for his complaint of religion.

In assessment Hawking sticks out now not simplest for his well-known explanations of science via television and the published word but for the substance of his thoughts.

His grandest idea issues the vital and everyday query every body has asked – how did the universe come into being?

Hawking defined to the arena, thru his artwork on black holes, how the entire cosmos changed into flung out of a singularity.

We could probably ask ourselves but, as critics have, how the technology expounded via Hawking has been conflated with movie megastar enchantment. Is it possible to critique his ideas with out critiquing the individual inside the again of those mind?

Critics remind us that his greatest thoughts, inspiring even though they may be and framed with the aid of extremely good calculations, are in huge component intuitive hypotheses, yet unproven.

Hawking Radiation remains unproven, and scientists have mentioned large problems even in his concept of the Big Bang.

First there may be the trouble of in which the trouble dormant within the singularity got here from. Hawking hopes to demonstrate sooner or later in the destiny how some factor can come from not anything, however for now it remains a hassle.

Some researchers difficulty out that the Big Bang seems to violate the Law of Entropy, which says that rely variety becomes less organized over time. Yet if this is so how do the celebrities and galaxies look like so particularly complex and prepared?

There are, in the long run, scientific alternatives to the Big Bang concept. Some scientists postulate a Big Bounce rather than a Big Bang. In this model our universe is one in every of an limitless collection of contracting and increasing universes, nicely maintaining off the hassle of introduction from not something.

There are high-quality theories as nicely. And however the Big Bang remains the primary, and for masses the handiest, identified principle of the universe. Few people aside from physicists seem capable of absolutely maintain close the idea, and but it has gripped the arena.

Are we being attentive to the man in preference to the idea? Has Hawking become a few form of demagogue, the excessive priest of generation?

We are reminded too that Hawking has been wrong on a number of sports and has freely admitted this.

And then we have to do not forget the elephant within the room. Stephen Hawking is someone who lives in his personal thoughts. His thoughts is the prisoner of a frame that no longer obeys it. Or to location it a few other manner, his thoughts has been freed from the distractions of the body and the needs of the flesh.

So has Hawking emerge as for us a form of scientific saint? Is he a highbrow ascetic, transported via the denial of the flesh to a few higher aircraft we mere mortals can simplest dream of?

Doubtless Hawking himself may recoil at the concept, however likely we are projecting our very own conceptions of genius and disability upon him, and unfairly so.

Hawking has in no manner preferred to be visible as disabled, however in all likelihood the notion of genius in him permits us to border him inner our personal fallacious, romanticized thoughts of what disability approach.

Perhaps unconsciously we are stimulated with the useful resource of the romantic concept of the savant, the individual confined in a single detail, yet gifted by using nature in every other via way of repayment.

Hawking should simply now not thank us for this again-passed reward, nor take delivery of as authentic with the good judgment of that concept. He does no longer agree with there may be justice inherent in the universe. There is handiest the inevitable float of direction and impact, without motive or cause.

Hawking now not often talks approximately himself, and almost in no way speaks about his emotions. However he did speak on the connection amongst his contamination and his paintings.

'When you're confronted with the opportunity of early demise,' he said, 'it makes you apprehend lifestyles's definitely worth living, and there are masses of factors you need to do.'

He isn't looking to beat death, but rather to stay every day as actually as he can, that is awesome recommendation for everybody.

He says he does not play to the media, nor use it to spruce his mind. The motive of his media appearances and books (he even wrote a series for kids together along with his daughter Lucy) is in reality to offer an cause of his paintings to the general public, something he considers crucial.

He admits that his research received't have an effect at the each day lives of human beings very an lousy lot at all. Black holes, string concept and quantum mechanics acquired't treatment ailments or surrender famine or supply global peace, 'but it's miles essential to apprehend wherein we come from and what we can expect to discover as we discover.'

It does now not seem that Hawking is a theoretical physicist for the recognition and glory. He does what he does because of the reality he loves it.

So may additionally want to it is that he does now not want to percentage his genius with the world? He doesn't want to reveal a few difficulty. He's simplest someone using his capabilities and having a superb time doing it.

The Theory of Everything and Other Media Portrayals

In 2014, films have been released in the United Kingdom which, in the beginning appearance, regarded not in all likelihood to make any shape of impact at the cinema going public.

The reason for this changed into their problem rely – in both instances, they focused on the lives of proper British Heroes.

But very an entire lot not of the James Bond type. No, the problem of these films grow to be intellectual geniuses who had made their mark at the lifestyle and information of the u. S. A..

That the lead actors inside the ones movies were of the remarkable younger sparks of English acting brought to the enchantment.

And that they'd attended widely known, competing public schools further raised the anticipation.

The comparisons between Eddie Redmayne gambling Stephen Hawking in 'The Theory of Everything' and Benedict Cumberbatch's portrayal of Alan Turing in 'The Imitation Game' have come to be the mentioned detail in British cinema.

Redmayne is an vintage Etonian. Whilst there, he had as quickly as played the lower back cease of an Elephant in a house play.

By evaluation, Cumberbatch is an antique Harrovian. The faculties are competition. Geographically they will be quite near — Harrow being in North London, Eton certainly to the West of Heathrow.

They compete over their alumni. Eton should probable throw former Prime Minister David Cameron but Harrow can trump with Winston Churchill (despite the fact that, reputedly, he hated the college).

Eton counters with the Royal Princes, Harrow check friends with…Winston Churchill.

Turing, the mission of Benedict Cumberbatch's film, come to be the mathematical genius who solved the enigma code.

This became the tool Germans used at some degree within the Second World War to deliver out messages to its military operators, and from which they might release assaults on convoys crossing the Atlantic.

Turing worked at Bletchley Park, a converted manor house in the center of England from wherein code breakers operated in a few unspecified time inside the destiny of the war.

He created the gadget which modified into able to interpret the German Enigma codes, remedy them and, many ought to argue, lead the allies to victory inside the struggle.

'The Imitation Game' is a super movie, but it emerge as trumped through the no longer possibly achievement of a movie approximately a disabled scientist.

Eddie Redmayne described the demanding situations of playing Stephen Hawking, for which he obtained an Oscar.

Firstly, his burgeoning admiration for the person offered its very personal troubles. He had spent six months gaining knowledge of into Hawking, and have become step by step to appearance the scientist as an idol.

So when they met for the primary time, he spewed a worried barrage of verbal exchange at his mission, his tension and delight at assembly his new hero combining to make him sincerely unintelligible.

During the research segment, Redmayne come to be fairly thorough.

He visited patients of amyotrophic lateral sclerosis (motor-neurone illness). In truth, he spent 4 months at the National Hospital for Neurology and Neurosurgery's hospital studying approximately the condition.

He talked to Stephen's circle of relatives. He have a have a look at (without entire comprehension) 'A Brief History of Time'. He watched pix of his state of affairs.

The film was to be about Hawking's university days, while he every met his companion and notion, Jane Wilde, and whilst he gotten smaller the lifestyles destroying clinical state of affairs.

He had preferred to be as prepared as viable earlier than assembly the first rate guy, however now Redmayne turned into about to be delivered to the character he is probably playing, he started out out out to have doubts.

What if the studies he had undertaken had given him a false impression of the scientist?

But, as quickly as he had calmed himself, Redmayne discovered an entire lot from the meeting.

He observed that Stephen's voice had come to be exceedingly slurry during the onset of his ailment.

The supplied a trouble that he and the director, James Marsh, had now not truely considered. They were left with a conundrum.

On the most effective hand, they wished to represent Hawking and his situation in as sincere a moderate as viable, but similarly they did have to produce some thing their aim market may moreover want to understand.

It is all very well being true to the story, but if the audience can't apprehend it, then that story isn't being knowledgeable.

Subtitles were a opportunity stated. But both director and actor felt that those may want to detract from, in preference to make easy, the story they wished to inform.

Chapter 5: The Ensuing Speech Brought Lots To The Power Of The Story

Another detail that Eddie Redmayne had not absolutely grasped was the humor and power of the scientist.

Although physical he's able to use great a very small amount of muscle tissue, the cheeky, active and offbeat humor of the 11 365 days vintage starting out at St Albans School decrease decrease back within the Fifties remains present.

It have become the addition of this that enabled the tale to be a effective tug on the feelings, in preference to a maudlin story.

Redmayne labored with a dance instructor to train his body that allows you to undertake the uncomfortable positions into which Stephen's frame has contorted.

It was important to try this first. After all, Redmayne can also be portraying a person in love, and it modified into critical that the physical factors did now not detract from his focus in this.

He located out that one difficulty suffers from the situation do is to pay attention as an entire lot into their few last active muscle tissues as feasible.

These end up conduits for expressing their feelings, and showing themselves.

With Stephen Hawking, it became his face, and in particular his eyebrows, that persisted to expose the actual man.

Redmayne worked hard, analyzing pix drastically, to try to seize how those commonly insignificant facial competencies worked.

However, 'The Theory of Everything' isn't always the great occasion wherein Stephen Hawking has appeared on movie.

Not with the resource of way of a protracted chalk.

It is a part of his appeal that he's so media pleasant. After all, maximum human beings might probable in all likelihood conflict to call each one-of-a-kind living scientists, but Stephen's name might be first on the lips at some stage in any quiz.

Perhaps considered certainly one of his most famous roles has been in TV's hit caricature comedy, 'The Simpsons'.

Stephen's daughter, Lucy, is privy to one of the scriptwriters on the US display, and determined that they preferred to write down an episode imparting the scientist.

She persuaded her father to participate. He voiced himself to his cartoon presence a number of instances.

He regarded within the season 10 episode, 'They Saved Lisa's Brain', which grow to be extensively utilized in archive pics in a season eleven episode.

He additionally took element in severa more performances. These embody the season 16 episode 'Don't Fear the Roofer'; season 18's 'Stop or My Dog Will Shoot!' and the season 22 show 'Elementary School Musical'.

Stephen moreover appeared in a documentary particular – 'The Simpsons: A Culture Show'.

Given the sturdiness of the display, and the choice of worldwide famend figures from well-known manner of life to be involved, it is some other first-rate success that Stephen seems within the top 20 of IGN's 'Top 25 Simpsons Guest Stars'.

He rolls in at range 16 – quite incredible for a cosmologist.

Stephen wrote the 2010 mini-series 'Into the Universe with Stephen Hawking'. This collection became created for the invention channel. It aired both in the US and in Britain.

However, in Britain, it become called 'Stephen Hawking's Universe.'

With a neat contact of irony, voice overs have been recorded thru the actor Benedict Cumberbatch, who ought to of path be the lead inside the rival movie to Eddie Redmayne's 'The Theory of Everything' in a few years' time.

In the technological know-how series, Hawking appears in character in each of the episodes – 'Aliens', 'Time Trael' and 'The Story of Everything'.

In the remaining of these, some photos turned into used for an episode of each exclusive documentary, 'Curiosity'.

'Star Trek' is, as all people apprehend, one of the terrific science fiction collection of all time, and on this Stephen Hawking created a 'first'.

In the episode 'Descent, Part 1' – the denouement of Season Six of 'Star Trek: The Next Generation' he appears as a hologram.

He is proven gambling poker with certainly one of a type quite recognizable names from the sector of era, Sir Isaac Newton and Albert Einstein.

However, the awesome issue about this is that he completed the hologram himself. In doing so he have become the primary ever visitor on 'Star Trek' to play himself.

Just as with 'The Simpsons', and the ironic portrayal of card playing on 'Star Trek', comedy is considered one of Stephen's desired genres for movie and TV.

Not unexpected, probably, given his usually humorous way.

He regarded in the British TV comedy legend 'Red Dwarf'. Stephen is a fan of the gathering, and has praised the writers for his or her witty use of faux-era.

He has moreover taken aspect in the big US technological know-how based comedy show 'The Big Bang Theory', which stars Jim Parsons.

He has seemed in no fewer than six episodes, regularly as a voice over. The episodes are:

'The Hawking Excitation', The Extract Obliteration', 'The Relationship Diremption', 'The Troll Manifestation', 'The Celebration Experimentation' and 'The Geology Elevation'.

Most years inside the United Kingdom, a big charity event is held referred to as Red Nose Day.

Created through comic Lenny Henry as a manner to address poverty in Africa, the charity is now part of British Culture.

It features severa zany activities taking area across the u . S . A ., and stars appear at the live display, in which people pledge cash.

Often in cross back for the ones stars doing a little factor funny, however undignified. Each yr, a modern layout of Red Nose is created, plus variations for your car.

With significant self-irony, Hawking regarded in 2015 in which he 'transforms' proper into a 'Transformer'.

The caricature featured cutting factor British comedians alongside Hawking — Catherine Tate and David Walliams.

He additionally took detail in a cartoon with Jim Carrey on 'Late Night with Conan O'Brien'. The left field, barely seditious 'TV Offal' featured him inside the pick out sequence.

Other appearances on TV include supplying the trophy for the winners of the hard quiz show, University Challenge.

He has been worried in severa technological knowledge based documentaries including 'Genius of Britain', in which he changed into link guy for a chain on well-known British Scientists.

He modified right right into a player within the documentary 'The eleventh Hour' and 'Alien Planet'. He also hosted, thru voice over, a generation fiction collection 'Masters of Science Fiction.'

Hawking's first reference to the actor Benedict Cumberbatch occurred sooner or later of the filming of the TV Movie 'Hawking'.

With unexpected similarity to the number one Hollywood movie some years later, this movie too portrayed Stephen's time Cambridge.

Cumberbatch played the eponymous lead within the movie.

Stephen changed into moreover portrayed with the aid of manner of an actor in an episode of 'Stargate Atlantis'. 'Brain Storm' and as nicely in 'Superhero Movie.'

On pinnacle of his numerous appearances in 'The Simpsons', he is likewise a regular in 'Futurama', having been a person in the following episodes.

In 'Anthology of Interest' he appears as a defend of the gap-time continuum. 'The

Beast with a Billion Backs' he has a simply curtailed function as he is proven as simply a head in a jar.

'Family Guy' is each different cool animated film display which has drawn cautiously on Hawking. He has regarded in seven episodes, most presently as a streaker in a basketball undertaking.

He is a online game person in the episode entitled 'reincarnation'.

Such is his recognizable voice, and so nicely and warmly is he regarded, that he is regularly mission to moderate parody, specifically in British comedy suggests.

A man or woman from the hit series 'The Vicar of Dibley', Frank (whose character is famend for being dull) chooses to portray his function as a Wise Man in a Nativity play as having Hawking's voice.

Keeping with the faux-spiritual challenge depend, he's referenced in the comedy 'Father Ted.'

Other times he has been referred to in a similar manner embody the American hit 'Seinfeld' among many others.

He has appeared considerably in commercials; embody a few for the excessive stop marque Jaguar.

Stephen moreover pre-recorded an fun skit providing his wheel chair for a Monty Python display.

In fact, his media appearances and representations are myriad. Amongst the extensive number now not indexed right here is the usage of his voice on a Pink Floyd album.

Perhaps one of his desired lampoonings concerned the satirical information e-book 'The Onion'.

In it the periodical runs a chunk of writing claiming that Hawking has designed a excessive powered robotic exoskeleton.

In preserving along together together with his love of a funny tale, Stephen wrote a letter published inside the paper wherein he claims that they have unveiled his plans for international domination.

It is high-quality, and of massive advantage, that someone so bodily afflicted may additionally need to come upon lifestyles with such positivity and experience of amusing.

Hawking is also a prolific writer. Some of his works have been referred to already on this e book, but in reality he has produced no fewer than twenty-six courses for which he's every co or sole creator.

This is in addition to the far too many to say articles he has written.

As a very last point for this biography, we are able to test a hard and fast of eight of these books which could likely nicely slip beneath the net.

Co-written alongside collectively along with his daughter Lucy, Stephen has produced this mini collection written specially for children.

The 'George' series feature George and his fantastic buddy Annie. The intrepid duo excursion into place, reading (as is the case with their readers as nicely) approximately scientific concepts as they go with the flow.

The recollections capture the off-beat, funny and innovative factor of Stephen which we observed proper yet again from a younger little one, while he should abandon Monopoly for video video video games of his personal, complicated, making.

An example of the creativity within the George books can be taken from their 2014 ebook 'George and the Unbreakable Code'.

In this splendid tale, we see the libertarian, anti-established order Hawking shine thru. Banks hand out free cash, supermarkets prevent being capable of fee for meals and aircraft refuse to fly.

All due to the truth exquisite pc systems seem to have been hacked.

In each exclusive in their books, 'George and the Big Bang', he demonstrates the respect for younger people so that you can usually win their adoration.

He explores taken into consideration one in all his maximum complicated theories however the children of his goal marketplace.

Here, George is having some domestic issues or even his notable pal Anne appears to provide different subjects on her thoughts.

So, he gadgets out to help every other mate, Eric, expand his plans to move lower again to the begin of the universe.

The story competencies evil doing villains and takes the shape of a story, series of essays or maybe a photo novel element.

'George's Secret Key to the Universe' explores physics, technological know-how and the universe as a whole through a exciting collection of adventures.

In this story, a extremely good laptop with its personal intelligence (called Cosmos) takes Annie, her father Eric and, of direction, George on an thrilling experience to the brink of a black hole – and optimistically domestic over again.

The principle of black holes described thru comedy and journey.

Brilliant.

Stephen Hawking isn't only a genius. He is an educator and someone with a razor sharp wit and huge revel in of a laugh.

Chapter 6: In His Very Non-Public Phrases

My advice to distinctive disabled humans will be, address subjects your disability does now not save you you doing properly, and do not remorse the matters it interferes with. Don't be disabled in spirit in addition to physical.

No one undertakes studies in physics with the intention of prevailing a prize. It is the delight of discovering some issue no man or woman knew in advance than.

Science is cute whilst it makes smooth reasons of phenomena or connections among special observations. Examples embody the double helix in biology and the essential equations of physics.

While physics and mathematics can also inform us how the universe started out, they are not masses use in predicting human behavior because there are an extended manner too many equations to treatment. I'm no higher than everyone else at expertise what makes human beings tick, in particular ladies.

I regard the mind as a laptop so that you can save you operating while its components fail. There isn't any heaven or afterlife for damaged down computer structures; that may be a fairy tale for human beings frightened of the darkish.

I turn out to be in no way pinnacle of the beauty at college, however my classmates ought to have seen potential in me, due to the fact my nickname have become 'Einstein.'

It is usually recognized that girls are higher than guys at languages, private contributors of the own family and multi-tasking, however an awful lot a good deal much less correct at map-studying and spatial reputation. It is therefore now not unreasonable to anticipate that girls might be a lot tons much less accurate at arithmetic and physics.

A few years in the beyond, the metropolis council of Monza, Italy, barred doggy proprietors from keeping goldfish in curved bowls... Pronouncing that it's miles merciless to hold a fish in a bowl with curved components because, searching at out, the fish may have a distorted view of truth. But how can we recognize we have had been given the proper, undistorted image of reality?

Obviously, because of my incapacity, I want help. But I even have typically tried to overcome the restrictions of my state of affairs and lead as complete a life as viable. I without a doubt have travelled the world, from the Antarctic to zero gravity.

God might also exist, however technological knowledge can give an reason behind the universe with out the want for a creator.

In a whole lot much less than 100 years, we've got got were given observed a contemporary manner to don't forget ourselves. From sitting on the center of the universe, we now find out ourselves orbiting a mean-sized solar, that is simply sincerely one in all hundreds of masses of stars in our personal Milky Way galaxy. Stephen Hawking

The human race may be the only smart beings inside the galaxy.

I expect computer viruses need to depend as existence. I anticipate it says something about human nature that the pleasant form of lifestyles we have got created to date is only negative. We've created existence in our very very own photograph.

If the charge of boom one 2nd after the Big Bang were smaller by using even one part in a hundred thousand million, it might have re-collapsed earlier than it reached its present duration. On the other hand, if it had been greater by using a element in one million, the universe could have progressed too unexpectedly for stars and planets to shape.

I do no longer have plenty exceptional to say approximately motor neuron ailment, but it taught me not to pity myself due to the reality others have been worse off, and to get on with what I still must do. I'm happier now than in advance than I developed the condition.

We are absolutely a complex breed of monkeys on a minor planet of a very common movie megastar. But we are able to apprehend the Universe. That makes us a few factor very special.

The widespread technique of era of building a mathematical model cannot solution the questions of why there need to be a universe for the version to offer an reason behind. Why does the universe visit all the trouble of current?

God no longer simplest plays cube, He also now and again throws the dice wherein they cannot be seen.

I consider everybody need to have a massive image of approaches the universe operates and our place in it. It is a fundamental human desire. And it additionally locations our issues in perspective. Stephen Hawking

Science is more and more answering questions that turned into as quickly as the province of religion.

It's time to determine to finding the solution, to look for existence beyond Earth. Mankind has a deep want to find out, to examine, to understand. We moreover happen to be sociable creatures. It is critical for us to understand if we're on my own within the dark.

My discovery that black holes emit radiation raised immoderate troubles of consistency with the rest of physics. I have now resolved the ones problems, however the answer grew to emerge as out to be now not what I anticipated.

I can't cover myself with a wig and dark glasses - the wheelchair offers me away.

If we do discover a entire idea, it need to be in time understandable in massive precept with the aid of way of everybody. Then we can all, philosophers, scientists, and in reality normal humans have the capacity to take part within the speak of why we and the universe exist.

I assume the thoughts is basically a computer and awareness is like a pc software. It will stop to run on the identical time as the computer is became off. Theoretically, it can be re-created on a neural community, however that might be very hard, as it'd require all one's memories.

I'm no longer fearful of lack of lifestyles, however I'm in no hurry to die. I certainly have plenty I need to do first.

The radiation left over from the Big Bang is just like that in your microwave oven but very a whole lot lots much less effective. It may want to warm temperature your pizza most effective to minus 271.Three*C - now not plenty perfect for defrosting the pizza, now not to mention cooking it. Stephen Hawking

I grow to be no longer a top notch pupil. I did now not spend a good buy time at university; I turn out to be too busy taking element in myself.

So lengthy because the universe had a starting, we ought to expect it had a creator. But if the universe is certainly completely self-contained, having no boundary or thing, it might have neither beginning nor surrender: it would absolutely be. What vicinity, then, for a author?

Before I out of place my voice, it become slurred, so handiest those close to me can also need to apprehend, but with the computer voice, I located I ought to supply famous lectures. I experience speaking technology. It is vital that the general public is familiar with simple technological understanding, if they're now not to go away essential selections to others.

We need to expand as brief as possible technology that make viable a proper away connection amongst mind and computer, in order that synthetic brains make contributions to human intelligence in preference to opposing it.

Even if there may be only one possible unified concept, it's far simply a difficult and fast of guidelines and equations. What is it that breathes fireside into the equations and makes a universe for them to offer an explanation for?

Chapter 7: Stephen Hawking

Professor Stephen William Hawking become born on 8th January 1942 (exactly 3 hundred years after the lack of existence of Galileo) in Oxford, England. His mother and father' house emerge as in north London however at some level within the second international struggle Oxford modified into taken into consideration a more secure area to have babies. When he was 8 his own family moved to St. Albans, a town about 20 miles north of London. At the age of 11, Stephen went to St. Albans School after which directly to University College, Oxford (1952); his father's old university. Stephen favored to have a look at mathematics in spite of the reality that his father ought to have preferred remedy. Mathematics became no longer available at University College, so he pursued physics as a substitute. After three years and not very a whole lot paintings, he have become provided a number one elegance honours degree in natural technological expertise.

In October 1962, Stephen arrived on the Department of Applied Mathematics and Theoretical Physics (DAMTP) on the University of Cambridge to do studies in cosmology, there being no-one going for walks in that location in Oxford at the time. His supervisor have come to be Dennis Sciama, notwithstanding the fact that he had was hoping to get Fred Hoyle who changed into going for walks in Cambridge. After gaining his PhD (1965) together along with his thesis titled 'Properties of Expanding Universes', he have become, first, a research fellow (1965) then Fellow for Distinction in Science (1969) at Gonville & Caius university. In 1966 he received the Adams Prize for his essay 'Singularities and the Geometry of Space-time'. Stephen moved to the Institute of Astronomy (1968), later transferring lower once more to DAMTP (1973), hired as a studies assistant, and posted his first educational e-book, The Large Scale Structure of Space-Time, with George Ellis. During the following few years, Stephen become elected a Fellow of the Royal Society (1974) and

Sherman Fairchild Distinguished Scholar at the California Institute of Technology (1974). He became a Reader in Gravitational Physics at DAMTP (1975), progressing to Professor of Gravitational Physics (1977). He then held the location of Lucasian Professor of Mathematics (1979-2009). The chair have become based in 1663 with money left in the will of the Reverend Henry Lucas who have been the Member of Parliament for the University. It turn out to be first held with the useful resource of Isaac Barrow and then in 1669 with the useful aid of Isaac Newton. Stephen is currently the Dennis Stanton Avery and Sally Tsui Wong-Avery Director of Research at DAMTP.

Professor Stephen Hawking has worked at the simple legal guidelines which govern the universe. With Roger Penrose he showed that Einstein's huge idea of relativity implied space and time would possibly have a beginning within the Big Bang and an result in black holes (1970). These outcomes indicated that it become crucial to unify modern relativity with quantum precept, the alternative terrific clinical improvement of the number one half of the 20 th century. One cease end result of this kind of unification that he determined grow to be that black holes ought to now not be truely black, however as an alternative need to emit 'Hawking' radiation and subsequently evaporate and disappear (1974). Another conjecture is that the universe has no aspect or boundary in imaginary time. This might mean that the way the universe started out out became completely determined thru the prison guidelines of technological know-how. Recently Stephen has been going for walks with colleagues on a possible selection to the black hollow information paradox, in which debate centres across the conservation of facts.

His many publications consist of The Large Scale Structure of Spacetime with G F R Ellis, General Relativity: An Einstein Centenary Survey, with W Israel, and three hundred Years of Gravitation, with W Israel. Among the famous books Stephen Hawking has published are his exceptional dealer A Brief History of Time, Black Holes and Baby Universes and Other Essays, The Universe in a Nutshell, The Grand Design and My Brief History.

Professor Stephen Hawking has 13 honorary levels. He grow to be provided CBE (1982), Companion of Honour (1989) and the Presidential Medal of Freedom (2009). He is the recipient of many awards, medals and prizes, most drastically the Fundamental Physics prize (2013), Copley Medal (2006) and the Wolf Foundation prize (1988). He is a Fellow of the Royal Society and a member of the USA National Academy of Sciences and the Pontifical Academy of Sciences.

In 1963 Stephen have emerge as diagnosed with ALS, a shape of Motor Neurone Disease, unexpectedly after his 21st birthday. In spite of being wheelchair-positive and depending on a computerised voice device for conversation Stephen continues to mix circle of relatives life (he has 3 youngsters and three grandchildren) collectively along with his studies into theoretical physics, similarly to an intensive programme of excursion and public lectures. He despite the fact that hopes to make it into area in the end.

Professor Stephen Hawking speaks approximately "Why We Should Go into Space" for the NASA Lecture Series, April 21, 2008.

Credit: NASA/Paul Alers

Stephen Hawking grow to be regarded as one of the most incredible theoretical physicists in facts. His art work at the origins and form of the universe, from the Big Bang to black holes, revolutionized the area, even as his fine-promoting books have appealed to readers who may not have Hawking's medical records. Hawking died on March 13, 2018.

Stephen Hawking, Famed Physicist Who Defied ALS Odds, Dies at 76

Stephen Hawking Remembered via Neil deGrasse Tyson and More on Twitter

Stephen Hawking's Most Intriguing Quotes at the Future of Humanity, Aliens and Women

Stephen Hawking's Most Far-Out Ideas About Black Holes

In this quick biography, we test Hawking's training and profession — beginning from his discoveries to the famous books he is written — and the disease that robbed him of mobility and speech.

British cosmologist Stephen William Hawking have become born in England on Jan. 8, 1942 — three hundred years to the day after the death of the astronomer Galileo Galilei. He attended University College, Oxford, in which he studied physics, notwithstanding his father's urging to interest on medicinal drug. Hawking went at once to Cambridge to analyze cosmology, the examine of the universe as an entire.

In early 1963, genuinely shy of his twenty first birthday, Hawking became identified with motor neuron illness, greater usually called Lou Gehrig's disorder or amyotrophic lateral sclerosis (ALS). He became not predicted to live more than years. Completing his doctorate did not appear in all likelihood. Yet, Hawking defied the odds, no longer pleasant accomplishing his Ph.D. But moreover forging new roads into the know-how of the universe in the a few years when you consider that.

As the disease spread, Hawking became less cell and started the use of a wheelchair. Talking grew extra difficult and, in 1985, an emergency tracheotomy brought on his overall loss of speech. A speech-generating tool constructed at Cambridge, mixed with a software program software software program application, served as his electronic voice, permitting Hawking to pick out his terms via shifting the muscle groups in his cheek.

Just earlier than his evaluation, Hawking met Jane Wilde, and the 2 have been married in

1965. The couple had three kids in advance than isolating. Hawking remarried in 1995 but divorced in 2006.

A remarkable mind

Hawking persisted at Cambridge after his commencement, serving as a research fellow and later as a professional fellow. In 1974, he changed into inducted into the Royal Society, a worldwide fellowship of scientists. In 1979, he changed into appointed Lucasian Professor of Mathematics at Cambridge, the maximum well-known instructional chair in the worldwide (the second holder became Sir Isaac Newton, moreover a member of the Royal Society.

Over the course of his profession, Hawking studied the fundamental felony pointers governing the universe. He proposed that, due to the truth the universe boasts a starting — the Big Bang — it probably may want to have an finishing. Working with fellow cosmologist Roger Penrose, he set up that Albert Einstein's Theory of General Relativity indicates that region and time started out out on the beginning of the universe and ends inside black holes, this means that that Einstein's idea and quantum concept want to be united.

Using the 2 theories collectively, Hawking additionally decided that black holes are not simply silent however instead emit radiation. He predicted that, following the Big Bang, black holes as tiny as protons have been created, dominated by the use of way of every elegant relativity and quantum mechanics. [PHOTOS: Black Holes of the Universe]

Professor Stephen Hawking reports the freedom of weightlessness during a 0 gravity flight.

Professor Stephen Hawking evaluations the liberty of weightlessness all through a zero gravity flight.

Credit: ZERO-G

In 2014, Hawking revised his idea, even writing that " there aren't any black holes" — at least, inside the way that cosmologists historically apprehend them. His principle removed the lifestyles of an "event horizon," the point in which not anything can break out. Instead, he proposed that there may be an "apparent horizon" that would adjust in step with quantum adjustments inside the black hollow. But the idea remains controversial

Hawking moreover proposed that the universe itself has no boundary, much like the Earth. Although the planet is finite, one should journey round it (and thru the universe) infinitely, in no way encountering a wall that could be defined due to the reality the "prevent."

Hawking's books

Hawking become a well-known creator. His first ebook, "A Brief History of Time" (tenth anniversary model: Bantam, 1998) became first published in 1988 and have become an global great vendor. In it, Hawking aimed to speak questions about the delivery and death of the universe to the layperson.

Hawking went on to put in writing other nonfiction books geared towards nonscientists. These consist of "A Briefer History of Time," "The Universe in a Nutshell," "The Grand Design" and "On the Shoulders of Giants." [Related: 8 Shocking Things We Learned From Stephen Hawking's Book "Grand Design"]

He and his daughter, Lucy Hawking, furthermore created a fictional series of books for middle school children on the introduction of the universe, which includes "George and the Big Bang" (Simon & Schuster, 2012).

Hawking made numerous tv appearances, which encompass a playing hologram of himself on "Star Trek: The Next Generation" and a cameo at the tv show "Big Bang Theory." PBS furnished an educational miniseries titled "Stephen Hawking's

Universe," which probes the theories of the cosmologist.

In 2014, a movie based totally totally on Hawking's life changed into launched. Called "The Theory of Everything," the movie drew praise from Hawking, who stated it made him replicate on his very very own existence. "Although I'm notably disabled, I were a success in my scientific paintings," Hawking wrote on Facebook in November 2014. "I adventure widely and had been to Antarctica and Easter Island, down in a submarine and up on a 0-gravity flight. One day, I choice to enter area."

Chapter 8: Stephen Hawking Prices

Hawking's fees variety from first-rate to poetic to debatable. Among them:

"Even if there may be extremely good one possible unified idea, it's far really a set of policies and equations. What is it that breathes hearth into the equations and makes a universe for them to explain? The normal method of technological information of building a mathematical model can not solution the questions of why there need to be a universe for the model to give an explanation for. Why does the universe visit all the trouble of gift? "

"All of my life, I were inquisitive about the huge questions that face us, and function tried to locate clinical solutions to them. If, like me, you have got checked out the celebs, and attempted to make feel of what you notice, you too have began out out to wonder what makes the universe exist."

"Science predicts that many special forms of universe might be spontaneously created out of not some thing. It is an issue of chance which we're in."

"The whole records of science has been the gradual popularity that sports do no longer appear in an arbitrary way, however that they reflect a tremendous underlying order, which can also or might not be divinely stimulated. "

"We need to looking for the greatest value of our motion."

"The superb enemy of knowledge isn't always lack of understanding, it's far the illusion of understanding."

"Intelligence is the ability to conform to exchange."

"It is not clean that intelligence has any extended-time period survival fee. "

"One cannot in reality argue with a mathematical theorem."

"It is a waste of time to be irritated about my incapacity. One has to get on with existence and I have not completed badly. People won't have time for you if you are constantly indignant or complaining."

A listing of Hawking costs is probably incomplete without bringing up some of his more controversial statements.

He regularly said that human beings ought to go away Earth if we favored to stay on.

"It might be difficult enough to avoid disaster in the next hundred years, no longer to say the subsequent thousand or million...Our simplest risk of long-time period survival isn't to stay inward-searching on planet Earth, however to spread out into location." — August 2010

"[W]e ought to ... maintain to go into area for the future of humanity...I do no longer anticipate we can stay to inform the story every other 1,000 years with out escaping beyond our fragile planet." — November 2016

"We are on foot out of location and the simplest places to visit are distinct worlds. It is time to discover unique solar structures. Spreading out may be the satisfactory thing that saves us from ourselves. I am satisfied that humans want to move away Earth." — June 2017

He additionally said time excursion need to be viable, and that we need to discover vicinity for the romance of it.

"Time journey turned into as soon as belief of as truly technology fiction, however Einstein's favored idea of relativity lets in for the opportunity that we can also need to warp place-time masses that you could burst off in a rocket and return earlier than you place out. I become one of the first to write down approximately the conditions underneath which this could be viable. I showed it would require matter number with bad power density, which might not be available. Other scientists took courage from my paper and wrote similarly papers at the venture," he knowledgeable Parade in 2010.

"Science isn't always only a disciple of purpose, but, additionally, considered one in all romance and ardour."

The theoretical physicist modified into additionally involved that robots could not high-quality have an impact at the economic machine but additionally advise doom for humanity.

"The automation of factories has already decimated jobs in traditional manufacturing, and the upward push of artificial intelligence is probable to extend this procedure destruction deep into the center instructions, with best the most being involved, revolutionary or supervisory roles remaining," he wrote in a 2016 column in The Guardian.

"The development of whole artificial intelligence have to spell the give up of the human race," he informed the BBC in 2014. Hawking added, but, that AI advanced up to now has been useful. It's greater the self-replication capability that issues him. "It may take off on its very own, and re-layout itself at an ever-growing rate. Humans, who're constrained with the resource of gradual organic evolution, couldn't compete, and can be preceding."

"The genie is out of the bottle. I fear that AI might also moreover moreover update humans altogether," Hawking informed WIRED in November 2017.

An avowed atheist, Hawking moreover every now and then waded into the subject of religion.

"Because there may be a law together with gravity, the universe can and will create itself from not anything. Spontaneous creation is the reason there may be something as opposed to now not anything, why the universe exists, why we exist. It isn't always essential to invoke God to mild the blue touch paper and set the universe going." — The Grand Design, via Stephen Hawking and Leonard Mlodinow

"I regard the thoughts as a computer with a purpose to prevent working whilst its components fail...There is not any heaven or afterlife for damaged down computers; that may be a fairy tale for humans fearful of the darkish." — 2011 interview with The Guardian

"Before we apprehend technological data, it is herbal to trust that God created the universe. But now technological expertise offers a greater convincing rationalization. What I meant with the useful resource of 'we'd recognize the thoughts of God' is, we'd apprehend the whole thing that God may want to recognize, if there have been a God, which there isn't always. I'm an atheist." — 2014 interview in El Mundo

Public Lectures

Included below are a selection of Professor Hawking's public lectures.

Into a Black Hole 2008

Is it possible to fall in a black hollow, and pop out in every other universe? Can you get away from a black hole when you fall interior? In this lecture I communicate approximately some of the topics I've located out approximately black holes.

The Origin of the Universe 2005

Why are we right proper right here? Where did we come from? The solution typically given turn out to be that humans have been of comparatively brand new starting, because it need to were obvious, even at early times, that the human race became improving in records and generation. So it can not had been round that long, or it would have superior even greater.

Godel and the End of Physics 2002

In this communicate, I want to ask how a protracted way can we skip in our look for expertise and understanding. Will we ever find an entire form of the legal hints of nature? By a complete shape, I imply a tough and rapid of regulations that during precept as a minimum allow us to anticipate the future to an arbitrary accuracy, expertise the u . S . A . Of the universe at one time. A qualitative statistics of the legal suggestions has been the motive of philosophers and scientists, from Aristotle onwards.

Space and Time Warps 1999

In technological information fiction, vicinity and time warps are a commonplace. They are used for fast trips across the galaxy, or for journey through time. But present day generation fiction, is regularly the next day's technology reality. So what are the probabilities for region and time warps.

Does God Play Dice 1999

This lecture is prepared whether or not or not we're capable of assume the future, or whether or now not or not it's miles arbitrary and random. In ancient times, the world must have appeared quite arbitrary. Disasters at the side of floods or illnesses ought to have appeared to manifest with out caution or apparent motive. Primitive human beings attributed such natural phenomena, to a pantheon of gods and goddesses, who behaved in a capricious and kooky manner. There turn out to be no manner to are watching for what they might do, and the quality preference modified into to win favour with the useful resource of manner of gives or moves.

Chapter 9: The Beginning Of Time 1996

In this lecture, I would really like to talk about whether or not time itself has a beginning, and whether it will have an surrender. All the proof seems to suggest, that the universe has now not existed for all time, however that it had a starting, about 15 billion years in the beyond. This might be the most exceptional discovery of contemporary cosmology. Yet it's far now taken as a proper. We aren't but sure whether or not or not the universe could have an surrender.

Life in the Universe 1996

In this talk, I would really like to invest a touch, on the development of existence inside the universe, and specifically, the development of sensible lifestyles. I shall take this to encompass the human race, no matter the truth that an entire lot of its behaviour through out records, has been pretty stupid, and now not calculated to useful useful useful resource the survival of the species.

Publications

A Smooth Exit from Eternal Inflation. S.W. Hawking, T. Hertog. 24 Jul 2017. 14pp, arXiv:1707.07702 [hep-th]

The Conformal BMS Group. S.J. Haco, S.W.Hawking, M.J.Perry, J.L.Bourjaily. 27 Jan 2017. 16pp, arXiv:1701.08110 [hep-th]

Superrotation Charge and Supertranslation Hair on Black Holes. S.W. Hawking, M.J. Perry, A. Strominger. High Energ. Phys. (2017) 2017: 161. DOI: 10.1007/JHEP05(2017)161.

Black holes: The Reith Lectures. S.W. Hawking, May 2016, ISBN-thirteen: 978-0857503572

George and the Blue Moon. L. Hawking, S.W.Hawking, Mar 2016, ISBN-13: 978-0857533272

Soft Hair on Black Holes. S.W. Hawking, M.J. Perry, A. Strominger. Jan five, 2016. 9pp. Published in Phys.Rev.Lett. 116 (2016) no.23, 231301, arXiv:1601.00921, DOI: 10.1103/PhysRevLett.116.231301

The Information Paradox for Black Holes, S.W. Hawking. Sep three, 2015. Three pp. DAMTP-2015-forty nine

George and the Unbreakable Code. L. Hawking. S.W.Hawking. Jun 2014. ISBN-thirteen: 978-0857533258

Information Preservation and Weather Forecasting for Black Holes. S.W. Hawking. Jan 2014. ArXiv:1401.5761v1 [hep-th]

My Brief History. S.W. Hawking, Sep 2013. A hundred and forty four pp ISBN-13: 978-0345535283

Vector Fields in Holographic Cosmology. James B.Hartle. S.W. Hawking, Thomas Hertog. May 2013. 17 pp. ArXiv:1305.7190v1 [hep-th], DOI: 10.1007/JHEP11(2013)201

George and the Big Bang. L. Hawking. S.W.Hawking. Aug 2012. ISBN-thirteen: 978-1442440050

Quantum Probabilities for Inflation from Holography. James B.Hartle. S.W. Hawking,

Thomas Hertog. Jul 2012. ArXiv:1207.6653v3 [hep-th], DOI: 10.1088/1475-7516/2014/01/half of

Accelerated Expansion from Negative Lambda. James B. Hartle (UC, Santa Barbara), S.W. Hawking (Cambridge U., DAMTP), Thomas Hertog (Leuven U. & Intl. Solvay Inst., Brussels). May 2012. 28 pp. ArXiv:1205.3807v3 [hep-th]

George's Cosmic Treasure Hunt. L. Hawking. S.W.Hawking. May 2011. ISBN-thirteen: 978-1442421752

The dreams that stuff is fabricated from: The maximum incredible papers of quantum physics - and the way they shook the scientific international. Stephen Hawking, (ed.) (Cambridge U., DAMTP). 2011. 1071 pp. Published in Philadelphia, USA: Running Pr. (2011) 1071 p. ISBN-thirteen: 978-0762434343

Local Observation in Eternal inflation. James Hartle (UC, Santa Barbara), S.W. Hawking

(Cambridge U.,DAMTP), Thomas Hertog (APC, Paris & Intl. Solvay Inst., Brussels). Sep 2010. Four pp. Published in Phys.Rev.Lett. 106 (2011) 141302. ArXiv:1009.2525v2 [hep-th], DOI: 10.1103/PhysRevLett.106.141302

The Grand Design. S.W.Hawking and L. Mlodinov (Sep 2010), ISBN-13: 978-0553805376

The No-Boundary Measure in the Regime of Eternal Inflation. James Hartle (UC, Santa Barbara), S.W. Hawking (Cambridge U., DAMTP), Thomas Hertog (APC, Paris & Intl. Solvay Inst., Brussels). Jan 2010. 29 pp.Published in Phys.Rev. D82 (2010) 063510. ArXiv:1001.0262v1 [hep-th], DOI: 10.1103/PhysRevD.Eighty .063510

George's Secret Key to the Universe. L. Hawking. S.W.Hawking. May 2009. ISBN-thirteen: 978-1416985846

Why did the Universe Inflate? S.W. Hawking (Cambridge U., DAMTP). 2009. 7 pp. DOI: 10.1007/978-0-387-87499-9_10

The Classical Universes of the No-Boundary Quantum State. James B. Hartle (UC, Santa Barbara), S.W. Hawking (Cambridge U.,DAMTP), Thomas Hertog (APC, Paris & Intl. Solvay Inst., Brussels). Mar 2008. Forty six pp. ArXiv:0803.1663 [hep-th], DOI: 10.1103/PhysRevD.Seventy seven.123537

No-Boundary Measure of the Universe. James B. Hartle (UC, Santa Barbara), S.W. Hawking (Cambridge U., DAMTP), Thomas Hertog (APC, Paris & Intl. Solvay Inst., Brussels). Nov 2007. Four pp. Published in Phys.Rev.Lett. A hundred (2008) 201301. ArXiv:0711.4630 [hep-th], DOI: 10.1103/PhysRevLett.A hundred.201301

Volume Weighting in the No Boundary Proposal. S.W. Hawking. Oct 2007. 7 pp. ArXiv:0710.2029 [hep-th]

God created the Integers. S.W.Hawking. Oct 2007. ISBN-thirteen: 978-0762430048

The degree of the universe. S.W. Hawking (Cambridge U., DAMTP). 2007. 6 pp.

Published in AIP Conf.Proc. 957 (2007) seventy nine-eighty 4, DOI: 10.1063/1.2823830

Populating the landscape: A Top down method. S.W. Hawking (Cambridge U., DAMTP), Thomas Hertog (CERN). CERN-PH-TH-2006-022. Feb 2006. 22 pp. Published in Phys.Rev. D73 (2006) 123527. ArXiv:hep-th/0602091, DOI: 10.1103/PhysRevD.Seventy three.123527

Information loss in black holes. S.W. Hawking (Cambridge U., DAMTP). DAMTP-2005-66. Jul 2005. Five pp.Published in Phys.Rev. D72 (2005) 084013. ArXiv:hep-th/0507171, DOI: 10.1103/PhysRevD.Seventy .084013

A non singular universe. S. Hawking (Cambridge U., DAMTP). 2005. 2 pp. Published in Phys.Scripta T117 (2005) 40 9-50

A briefer statistics of time. S. Hawking (Cambridge U., DAMTP), L. Mlodinow. 2005. 189 pp. Published in Reinbek, Germany:

Rowohlt (2005) 189 p. ISBN-thirteen: 978-0553385465

Black holes and the statistics paradox. S. Hawking (Cambridge U., DAMTP). Jul 2004. 7 pp. Prepared for Conference: C04-07-18, p.Fifty six-sixty Proceedings

The grand Stephen Hawking reader: Life and paintings. H. Mania, (ed.), S. Hawking. 2004. 291 pp. Published in (rororo. 61655)

Cosmology from the top down. Stephen W. Hawking (Cambridge U., DAMTP). DAVISINFLATION-2003-PELLY. May 2003. Published in In *Carr, Bernard (ed.): Universe or multiverse?* 91-98. ArXiv:astro-ph/0305562

On the Shoulders of Giants. N. Copernicus, J. Kepler, G. Galalei, I. Newton, A. Einstein (Author), S. Hawking. Dec 2003. ISBN-thirteen: 978-0762416981

The illustrated principle of the whole thing: The foundation and destiny of the universe. S.W. Hawking (Cambridge U., DAMTP). 2003.

119 pp. Published in Beverly Hills, USA: New Millennium Pr. (2003) 119 p

Brane new global.Stephen Hawking (Cambridge U., DAMTP). Aug 2002. 7 pp. Published in Conf.Proc. C0208124 (2002) 1-7

Why does inflation start at the pinnacle of the hill? S.W Hawking, Thomas Hertog (Cambridge U., DAMTP). Apr 2002. 21 pp. Published in Phys.Rev. D66 (2002) 123509. ArXiv:hep-th/0204212, DOI: 10.1103/PhysRevD.Sixty six.123509

Sixty years in a nutshell. S. Hawking (Newton Inst. Math. Sci., Cambridge). Jan 2002. Prepared for Workshop on Conference at the Future of Conference: C02-01-07.7

Chronology safety: Making the area stable for historians. S.W. Hawking. 2002. Published in In *Hawking, S.W. Et al.: The future of spacetime* 87-108

The Future of space-time. S.W. Hawking, K.S. Thorne, I. Novikov, T. Ferris, A. Lightman, R.

Price. 2002. 220 pp.Published in New York, USA: Norton (2002) 220 p

Why does inflation start on the top of the hill? S.W. Hawking (Cambridge U., DAMTP). Nov 2001. Prepared for Conference: C01-11-thirteen.1

Living with ghosts. S.W. Hawking, Thomas Hertog (Cambridge U., DAMTP). Jul 2001. Thirteen pp. Published in Phys.Rev. D65 (2002) 103515. ArXiv:hep-th/0107088, DOI: 10.1103/PhysRevD.Sixty 5.103515

The universe in a nutshell. S. Hawking (Cambridge U., DAMTP). 2001. 224 pp. ISBN-thirteen: 978-0553802023

Trace anomaly pushed inflation. S.W. Hawking, T. Hertog (Cambridge U., DAMTP), H.S. Reall (Queen Mary, U. Of London). DAMTP-2000-ninety two, QMW-PH-00-10. Oct 2000. Forty pp. Published in Phys.Rev. D63 (2001) 083504. ArXiv:hep-th/0010232, DOI: 10.1103/PhysRevD.Sixty 3.083504

Large N cosmology.S.W. Hawking (Cambridge U., DAMTP). Sep 2000. Prepared for Conference: C00-09-04.Four

Brane new international. S.W. Hawking, T. Hertog, H.S. Reall (Cambridge U., DAMTP). DAMTP-2000-25. Mar 2000. 28 pp.Published in Phys.Rev. D62 (2000) 043501. ArXiv:hep-th/0003052, DOI: 10.1103/PhysRevD.62.043501

Gravitational waves in open de Sitter area. S.W. Hawking, Thomas Hertog, Neil Turok (Cambridge U., DAMTP). Mar 2000. 17 pp. Published in Phys.Rev. D62 (2000) 063502. ArXiv:hep-th/0003016, DOI: 10.1103/PhysRevD.Sixty two.063502

de Sitter entropy, quantum entanglement and AdS / CFT. Stephen Hawking (Cambridge U., DAMTP), Juan Martin Maldacena, Andrew Strominger (Harvard U.). Feb 2000. 14 pp. Published in JHEP 0105 (2001) 001. ArXiv:hep-th/0002145, DOI: 10.1088/1126-6708/2001/05/001

Stability of AdS and section transitions. S.W. Hawking (Cambridge U., DAMTP). 2000. Published in Class.Quant.Grav. 17 (2000) 1093-1099, DOI: 10.1088/0264-9381/17/5/318

Brane worldwide black holes. A. Chamblin, S.W. Hawking, H.S. Reall (Cambridge U., DAMTP). DAMTP-1999-133. Sep 1999. Nine pp. Published in Phys.Rev. D61 (2000) 065007. ArXiv:hep-th/9909205, DOI: 10.1103/PhysRevD.Sixty one.065007

Charged and rotating AdS black holes and their CFT duals. S.W. Hawking, H.S. Reall (Cambridge U., DAMTP). DAMTP-R-99-108. Aug 1999. 18 pp. Published in Phys.Rev. D61 (2000) 024014. ArXiv:hep-th/9908109, DOI: 10.1103/PhysRevD.Sixty one.024014

Primordial black holes: Pair introduction, Lorentzian state of affairs, and evaporation. R. Bousso (Stanford U., Phys. Dept.), S.W. Hawking (Cambridge U.). 1999. Published in Int.J.Theor.Phys. 38 (1999) 1227-, DOI:10.1023/A:1026618832525

A debate on open inflation. S.W. Hawking (Cambridge U., DAMTP). Nov 1998. Published in AIP Conf.Proc. 478 (1999) 15-22

Rotation and the AdS / CFT correspondence. S.W. Hawking, C.J. Hunter, Marika Taylor (Cambridge U., DAMTP). Nov 1998. 24 pp. Published in Phys.Rev. D59 (1999) 064005. ArXiv:hep-th/9811056, DOI: 10.1103/PhysRevD.Fifty 9.064005

Nut charge, anti-de Sitter place and entropy. S.W. Hawking, C.J. Hunter (Cambridge U.), Don N. Page (Alberta U.). DAMTP-ninety eight-122. Sep 1998. Thirteen pp. Published in Phys.Rev. D59 (1999) 044033. ArXiv:hep-th/9809035, DOI: 10.1103/PhysRevD.Fifty nine.044033

Gravitational entropy and international structure. S.W. Hawking, C.J. Hunter (Cambridge U.). DAMTP-98-104. Aug 1998. 19 pp. Published in Phys.Rev. D59 (1999) 044025. ArXiv:hep-th/9808085, DOI: 10.1103/PhysRevD.Fifty 9.044025

Open inflation. S.W. Hawking (CAMBRIDGE U.). Aug 1998. Prepared for 2d Samos Meeting on Cosmology, Geometry and Re Conference: C98-08-31.Four

Lorentzian situation in quantum gravity. Raphael Bousso (Stanford U., Phys. Dept.), Stephen W. Hawking (Cambridge U.). SU-ITP-ninety eight-26, DAMTP-ninety eight-87. Jul 1998. 14 pp. Published in Phys.Rev. D59 (1999) 103501, Erratum-ibid. D60 (1999) 109903. ArXiv:hep-th/9807148, DOI: 10.1103/PhysRevD.60.109903, 10.1103/PhysRevD.Fifty nine.103501

Inflation, singular instantons and eleven-dimensional cosmology. S.W. Hawking, Harvey S. Reall (Cambridge U.). DAMTP-98-80 5. Jul 1998. 19 pp. Published in Phys.Rev. D59 (1999) 023502. ArXiv:hep-th/9807100, DOI: 10.1103/PhysRevD.Fifty nine.023502

Open inflation, the 4 form and the cosmological constant. Neil Turok, S.W. Hawking (Cambridge U.). Mar 1998. Eleven pp. Published in Phys.Lett. B432 (1998) 271-

278. ArXiv:hep-th/9803156, DOI: 10.1016/S0370-2693(98)00651-0

Open inflation with out fake vacua. S.W. Hawking, Neil Turok (Cambridge U.). Feb 1998. 10 pp. Published in Phys.Lett. B425 (1998) 25-32. ArXiv:hep-th/9802030, DOI: 10.1016/S0370-2693(ninety eight)00234-2

Comment on 'quantum introduction of an open universe', with the aid of Andrei Linde. S.W. Hawking, Neil Turok (Cambridge U.). Feb 1998. Four pp. ArXiv:gr-fine control/9802062

Is data misplaced in black holes?. S.W. Hawking (Cambridge U.). 1998. Published in In *Wald, R.M. (ed.): Black holes and relativistic stars* 221-240

Bulk costs in eleven-dimensions. S.W. Hawking, Marika Taylor (Cambridge U.). DAMTP-R-90 seven-fifty . Nov 1997. 26 pp. Published in Phys.Rev. D58 (1998) 025006. ArXiv:hep-th/9711042, DOI: 10.1103/PhysRevD.Fifty 8.025006

Evaporation of cosmological black holes. R. Bousso (Stanford U., Phys. Dept.), S.W. Hawking (Cambridge U., DAMTP). Nov 1997. 14 pp. Prepared for Conference: C97-eleven-11.1

(Anti)evaporation of Schwarzschild-de Sitter black holes. Raphael Bousso, Stephen W. Hawking (Cambridge U.). DAMTP-R-90 seven-26. Sep 1997. Sixteen pp. Published in Phys.Rev. D57 (1998) 2436-2442. ArXiv:hep-th/9709224, DOI: 10.1103/PhysRevD.Fifty seven.2436

Models for chronology choice. M.J. Cassidy, S.W. Hawking (Cambridge U.). DAMTP-R-ninety seven-forty seven. Sep 1997. 20 pp.Published in Phys.Rev. D57 (1998) 2372-2380. ArXiv:hep-th/9709066, DOI: 10.1103/PhysRevD.Fifty seven.2372

Evaporation of primordial black holes. S.W. Hawking (Cambridge U., DAMTP). Aug 1997. Prepared for 6th Conference on Quantum Mechanics of Conference: C97-eleven-eleven.1

Trace anomaly of dilaton coupled scalars in - dimensions. Raphael Bousso, Stephen W. Hawking (Cambridge U.). DAMTP-R-90 seven-25. May 1997. Eleven pp. Published in Phys.Rev. D56 (1997) 7788-7791. ArXiv:hep-th/9705236, DOI: 10.1103/PhysRevD.56.7788

Loss of quantum coherence thru scattering off virtual black holes. S.W. Hawking (Cambridge U.), Simon F. Ross (UC, Santa Barbara). DAMTP-R-90 seven-21, UCSB-TH-90 seven-08. May 1997. 29 pp. Published in Phys.Rev. D56 (1997) 6403-6415. ArXiv:hep-th/9705147, DOI: 10.1103/PhysRevD.Fifty six.6403

Evolution of near extremal black holes. S.W. Hawking, Marika Taylor (Cambridge U.). DAMTP-R-96-fifty six. Feb 1997. 25 pp. Published in Phys.Rev. D55 (1997) 7680-7692. ArXiv:hep-th/9702045, DOI: 10.1103/PhysRevD.Fifty five.7680

Evaporation of cosmological black holes. R. Bousso (Stanford U., Phys. Dept.), S.W. Hawking (Cambridge U., DAMTP). 1997.

Published in In *Ambleside 1997, Particle physics and the early universe* 481-494

The Nature of space and time. S.W. Hawking, R. Penrose. Jul 1996. Published in Sci.Am. 275 (1996) forty four-49

Pair advent of black holes throughout inflation. Raphael Bousso, Stephen W. Hawking (Cambridge U.). DAMTP-R-ninety six-33. Jun 1996. 29 pp. Published in Phys.Rev. D54 (1996) 6312-6322

Loss of facts in black holes. S. Hawking (Cambridge U., DAMTP). Jun 1996. Prepared for Conference on Geometric Issues in Foundations Conference: C96-06-25.2

Primordial black holes: Tunneling rather than no boundary notion. Raphael Bousso, Stephen W. Hawking (Cambridge U., DAMTP). DAMTP-R-ninety six-34, C96-05-25. May 1996. 14 pp. Published in Grav.Cosmol.Suppl. Four (1998) 28-37. ArXiv:gr-pleasant controls/9608009

Pair advent and evolution of black holes in inflation. Raphael Bousso, Stephen W.

Hawking (Cambridge U.). DAMTP-R-96-35, C96-05-26. May 1996. 8 pp. Published in Helv.Phys.Acta 69 (1996) 261-264. ArXiv:gr-excellent controls/9608008, DOI: 10.1103/PhysRevD.Fifty four.6312

The Gravitational Hamiltonian in the presence of nonorthogonal obstacles. S.W. Hawking, C.J. Hunter (Cambridge U.). DAMTP-R-ninety six-nine. Mar 1996. 19 pp. Published in Class.Quant.Grav. Thirteen (1996) 2735-2752. ArXiv:gr-quality manipulate/9603050, DOI: 10.1088/0264-9381/thirteen/10/012

Black holes in inflation. R. Bousso, S.W. Hawking (Cambridge U., DAMTP). 1996. Published in Nucl.Phys.Proc.Suppl. 57 (1997) 201-205, DOI: 10.1016/S0920-5632(90 seven)00377-0

The Nature of space and time. S. Hawking, R. Penrose. 1996. Published in Princeton, USA: Univ. Pr. (1996) 141 p. (The Isaac Newton Institute collection of lectures)

Virtual black holes. S.W. Hawking (Cambridge U.). DAMTP-R-ninety five-50. Oct 1995. 24 pp. Published in Phys.Rev. D53 (1996) 3099-3107. ArXiv:hep-th/9510029, DOI: 10.1103/PhysRevD.53.3099

The Probability for primordial black holes. R. Bousso, S.W. Hawking (Cambridge U.). DAMTP-R-90 5-33. Jun 1995. 15 pp. Published in Phys.Rev. D52 (1995) 5659-5664. ArXiv:gr-nice control/9506047, DOI: 10.1103/PhysRevD.Fifty .5659

Pair production of black holes on cosmic strings. S.W. Hawking, Simon F. Ross (Cambridge U.). DAMTP-R-90 five-30. Jun 1995. Nine pp. Published in Phys.Rev.Lett. Seventy 5 (1995) 3382-3385. ArXiv:gr-exceptional controls/9506020, DOI: 10.1103/PhysRevLett.75.3382

Duality among electric and magnetic black holes. S.W. Hawking, Simon F. Ross (Cambridge U.). DAMTP-R-ninety 5-8. Apr 1995. Sixteen pp. Published in Phys.Rev. D52

(1995) 5865-5876. ArXiv:hep-th/9504019, DOI: 10.1103/PhysRevD.52.5865

The Gravitational Hamiltonian, motion, entropy and floor phrases. S.W. Hawking (Cambridge U.), Gary T. Horowitz (UC, Santa Barbara). DAMTP-R-90 4-fifty two, UCSBTH-90 4-37. Jan 1995. Thirteen pp. Published in Class.Quant.Grav. Thirteen (1996) 1487-1498. ArXiv:gr-nice controls/9501014, DOI: 10.1088/0264-9381/thirteen/6/017

Quantum coherence and closed timelike curves. S.W. Hawking (Cambridge U.). DAMTP-R-ninety five-04. Jan 1995. 12 pp. Published in Phys.Rev. D52 (1995) 5681-5686. ArXiv:gr-nice controls/9502017, DOI: 10.1103/PhysRevD.Fifty two.5681

Black holes and child universes and other essays. S. Hawking. 1995. Published in Toronto, Canada: Bantam Books (1994) 172 p, ISBN-13: 978-0553374117

Entropy, Area, and black hole pairs. S.W. Hawking, Gary T. Horowitz (Newton Inst.

Math. Sci., Cambridge), Simon F. Ross (Cambridge U.). NI-94-012, DAMTP-R-90 four-26, UCSBTH-90 4-25. Sep 1994. 24 pp. Published in Phys.Rev. D51 (1995) 4302-4314. ArXiv:gr-excellent manage/9409013, DOI: 10.1103/PhysRevD.51.4302

Nature of area and time. S.W. Hawking (Cambridge U.). Sep 1994. Sixty pp. ArXiv:hep-th/9409195

Euclidean quantum gravity. G.W. Gibbons, (ed.), S.W. Hawking, (ed.) (Cambridge U.). 1994. Published in Singapore, Singapore: World Scientific (1993) 586 p

The Superscattering matrix for 2-dimensional black holes. S.W. Hawking (Cambridge U. & Caltech). Nov 1993. 12 pp. Published in Phys.Rev. D50 (1994) 3982-3986. ArXiv:hep-th/9401109, DOI: 10.1103/PhysRevD.50.3982

Quantum coherence in -dimensions. S.W. Hawking, J.D. Hayward (Cambridge U. & Caltech). CALT-sixty eight-1861, DAMTP-R-ninety 3-12. Mar 1993. 14 pp. Published in

Phys.Rev. D49 (1994) 5252-5256. ArXiv:hep-th/9305165, DOI: 10.1103/PhysRevD.49.5252

Supersymmetric Bianchi models and the rectangular root of the Wheeler-DeWitt equation. P.D. D'Eath, S.W. Hawking (Cambridge U.), O. Obregon (Guanajuato U., FIMEE). DAMTP-R-ninety -forty 4. Feb 23, 1993. Eleven pp. Published in Phys.Lett. B300 (1993) forty four-48, DOI: 10.1016/0370-2693(ninety 3)90746-5

The Origin of time asymmetry. S.W. Hawking (Cambridge U.), R. Laflamme (Cambridge U. & Los Alamos), G.W. Lyons (Cambridge U.). PRINT-90 three-0178 (DAMTP,CAMBRIDGE). Feb 12, 1993. Forty one pp. Published in Phys.Rev. D47 (1993) 5342-5356. ArXiv:gr-excellent control/9301017, DOI: 10.1103/PhysRevD.Forty seven.5342

Einstein's dream: Expeditions to the frontiers of area-time. Black holes and toddler universes and exclusive essays. (In German). S.W. Hawking. 1993. Published in Reinbek, Germany: Rowohlt (1993) one hundred 90 p

Naked and thunderbolt singularities in black hollow evaporation. S.W. Hawking, J.M. Stewart (Cambridge U.). PRINT-ninety -0362 (DAMTP,CAMBRIDGE), DAMTP-R-ninety -37. Jul 1992. 28 pp. Published in Nucl.Phys. B400 (1993) 393-415. ArXiv:hep-th/9207105, DOI: 10.1016/0550-3213(90 3)90410-Q

Evaporation of -dimensional black holes. S.W. Hawking (Caltech & Cambridge U.). CALT-sixty eight-1774. Mar 20, 1992. Eleven pp. Published in Phys.Rev.Lett. Sixty 9 (1992) 406-409. ArXiv:hep-th/9203052, DOI: 10.1103/PhysRevLett.Sixty nine.406

Kinks and topology alternate. G.W. Gibbons, S.W. Hawking (Cambridge U.). 1992. Published in Phys.Rev.Lett. Sixty 9 (1992) 1719-1721, DOI: 10.1103/PhysRevLett.Sixty nine.1719

Evaporation of two-dimensional black holes. S.W. Hawking (Cambridge U.). 1992. Published in In *Trieste 1992, Proceedings, The renaissance of fashionable relativity and cosmology* 274-286

Selection hints for topology alternate. G.W. Gibbons, S.W. Hawking (Cambridge U.). PRINT-91-0452 (DAMTP,CAMBRIDGE). Nov 12, 1991. 14 pp. Published in Commun.Math.Phys. 148 (1992) 345-352, DOI: 10.1007/BF02100864

The no boundary situation and the arrow of time. S.W. Hawking (Cambridge U., DAMTP). Sep 1991. Prepared for NATO Workshop on the Physical Origin of Conference: C91-09-30.Four

The Chronology safety conjecture. S.W. Hawking (Cambridge U.). DAMTP-R-91-15. Jul 1991. 24 pp.Published in Phys.Rev. D46 (1992) 603-611, DOI: 10.1103/PhysRevD.46.603

Wormholes in string idea. Alex Lyons (Alberta U.), S.W. Hawking (Cambridge U.). ALBERTA-THY-5-ninety one. May 1991. 40 pp. Published in Phys.Rev. D44 (1991) 3802-3818, DOI: 10.1103/PhysRevD.44.3802

The Alpha parameters of wormholes. S.W. Hawking (Cambridge U.). 1991. Published in

Phys.Scripta T36 (1991) 222-227, DOI: 10.1088/0031-8949/1991/T36/023

The chronology safety conjecture. S.W. Hawking (Cambridge U., DAMTP). 1991. Published in In *Kyoto 1991, Recent tendencies in theoretical and experimental popular relativity, gravitation and relativistic field theories, pt. A* 3-thirteen

Beginning or give up? Inaugural lecture. (In German). S. Hawking, (ed.) (Cambridge U.). 1991. Published in Paderborn, Germany: Junfermann (1991) 40 3 p

The Effective movement for wormholes. S.W. Hawking (Cambridge U.). PRINT-90-0682 (CAMBRIDGE). Nov 23, 1990. 19 pp. Published in Nucl.Phys. B363 (1991) 117-131, DOI: 10.1016/0550-3213(91)90237-R

The beginning of the universe. S.W. Hawking (Cambridge U., DAMTP). Sep 1990. Prepared for (IUPAP) International Conference on Primordial Conference: C90-09-04.2

The spectrum of wormholes. S.W. Hawking (Santa Barbara, KITP & Cambridge U.), Don N. Page (Santa Barbara, KITP & Penn State U. & Alberta U.). NSF-ITP-ninety-76. Jun 17, 1990. 37 pp. Published in Phys.Rev. D42 (1990) 2655-2663, DOI: 10.1103/PhysRevD.Forty .2655

Gravitational radiation from collapsing cosmic string loops. S.W. Hawking (Cambridge U.). DAMTP/R-90-14. Apr 1990. 7 pp. Published in Phys.Lett. B246 (1990) 36-38, DOI: 10.1016/0370-2693(90)91304-T

Wormholes and nonsimply connected manifolds. S.W. Hawking (Cambridge U.). DAMTP-R-ninety-thirteen. Jan 1990. 23 pp. Published in In *Jerusalem 1989, Proceedings, Quantum cosmology and baby universes* 245-267 and Cambridge Univ. - DAMTP-R-ninety-thirteen (ninety,rec.Jul.) 23 p

Baby universes. 2. S.W. Hawking (Cambridge U.). 1990. Published in Mod.Phys.Lett. A5 (1990) 453-466, DOI: 10.1142/S0217732390000524

Wormholes in dimensions 1 - 4. S.W. Hawking (Cambridge U.). 1990. Published in In *Boston 1990, Proceedings, Particles, strings and cosmology* 623-634. (see HIGH ENERGY PHYSICS INDEX 29 (1991) No.9950)

The Formation and evolution of cosmic strings. Proceedings, Workshop, Cambridge, UK, July three-7, 1989.G.W. Gibbons, (ed.), S.W. Hawking, (ed.), T. Vachaspati, (ed.) (Cambridge U. & Tufts U.). 1990. Published in Cambridge, UK: Univ. Pr. (1990) 542 p

Chapter 10: Do Wormholes Fix The Constants Of Nature?

S.W. Hawking (Cambridge U.). Print-89-0795 (CAMBRIDGE), DAMTP/R-89/thirteen. May 1989. 12 pp. Published in Nucl.Phys. B335 (1990) a hundred and fifty five, DOI: 10.1016/0550-3213(ninety)90175-D

The Edge Of Space-time. S. Hawking (Cambridge U.). 1989. Published in IN *DAVIES, P. (ED.): THE NEW PHYSICS* 61-69

Baby Universes And The Nonrenormalizability Of Gravity. S.W. Hawking, R. Laflamme (Cambridge U.). Print-88-0290(CAMBRIDGE), DAMTP/R-88/3. Mar 1988. 6 pp. Published in Phys.Lett. B209 (1988) 39, DOI: 10.1016/0370-2693(88)91825-4

Wormholes in Space-Time. S.W. Hawking (Cambridge U.). 1988. Published in Phys.Rev. D37 (1988) 904-910, DOI: 10.1103/PhysRevD.37.904

Baby universes. S.W. Hawking (Cambridge U.). 1988. Published in In *Leningrad 1988,

Proceedings, A.A. Friedmann:Centenary quantity* eighty one-ninety two.

A Brief History Of Time. S.W. Hawking. 1988. Published with the useful resource of Bantam (Sep 1988) 212p, ISBN-13: 978-0553380163

Quantum Cosmology. S.W. Hawking (Cambridge U.). 1988. Published in IN *FANG, LI-ZHI (ED.), RUFFINI, R. (ED.): QUANTUM COSMOLOGY*, one hundred ninety-235 AND PREPRINT - HAWKING, S.W. (80 3,REC.DEC.) sixty 4 P.

The Quantum Theory Of The Universe. S.W. Hawking (Cambridge U.). 1988. Published in IN *JERUSALEM 1983/84, PROCEEDINGS, INTERSECTION BETWEEN ELEMENTARY PARTICLE PHYSICS AND COSMOLOGY*, 71-90 seven.

Black Holes From Cosmic Strings. S.W. Hawking (Cambridge U.). Print-88-0310 (CAMBRIDGE). Dec 1987. Five pp.Published in Phys.Lett. B231 (1989) 237, DOI: 10.1016/0370-2693(89)90206-2

The Direction Of Time. S.W. Hawking (Cambridge U.). Print-87-0849 (DAMTP). Nov 10, 1987. 10 pp. Published in New Sci. A hundred and fifteen (1987) forty six

How in all likelihood is inflation?. S.W. Hawking (Cambridge U.), Don N. Page (Penn State U.). Print-87-0739 (PENN STATE). Jun 1987. 30 pp. Published in Nucl.Phys. B298 (1988) 789-809, DOI: 10.1016/0550-3213(88)90008-9

The Origin Of The Universe. S.W. Hawking (Cambridge U.). Print-87-0841 (CAMBRIDGE). Jun 1987. 10 pp.

The Ground State Of The Universe. S.W. Hawking (Cambridge U.). Print-87-0845 (CAMBRIDGE), C87/05/01.2. May 1987. 3 pp. Closing Remarks given at Conference: C87-05-01.2

Quantum Coherence Down the Wormhole. S.W. Hawking (Cambridge U.). Print-87-0842 (CAMBRIDGE). Apr 1987. 12 pp. Published in

Phys.Lett. B195 (1987) 337, DOI: 10.1016/0370-2693(87)90028-1

The Schrodinger Equation In Quantum Cosmology And String Theory. S.W. Hawking (Cambridge U.). Print-87-0843 (CAMBRIDGE). Mar 1987. 10 pp.

Three Hundred Years Of Gravitation. S.W. Hawking, (Ed.), W. Israel, (Ed.). 1987. Published in Cambridge, UK: Univ. Pr. (1987) 684 p

Quantum Cosmology. S.W. Hawking (Cambridge U.). Print-87-0166 (CAMBRIDGE), C87/06/29. Dec 1986. 31 pp.Published in In *Hawking, S.W. (ed.), Israel, W. (ed.): Three hundred years of gravitation*, 631-651 and Preprint - Hawking, S.W. (86,rec.Jan.87) 31 p

A Natural Measure On The Set Of All Universes. G.W. Gibbons, S.W. Hawking, J.M. Stewart (Cambridge U.).PRINT-86-1241. Oct 14, 1986. 19 pp. Published in Nucl.Phys. B281 (1987) 736, DOI: 10.1016/0550-3213(87)90425-1

The Density Matrix Of The Universe. S.W. Hawking (Cambridge U.). PRINT-86-0918 (CAMBRIDGE). Apr 1986. 10 pp. Published in Phys.Scripta T15 (1987) 151, DOI: 10.1088/0031-8949/1987/T15/020

Lectures On Quantum Cosmology. S.W. Hawking (Cambridge U.). 1986. Published in In *Kyoto 1985, Proceedings, Quantum Gravity and Cosmology*, a hundred seventy-206

Lectures On Quantum Cosmology S.W. Hawking (Cambridge U.). 1986.

Published in In *De Vega, H.J. (Ed.), Sanchez, N. (Ed.): Field Theory, Quantum Gravity and Strings*, 1-forty five

Who's Afraid Of (higher Derivative) Ghosts?. S.W. Hawking (Cambridge U.). Print-86-0124 (CAMBRIDGE). Sep 1985. 16 pp. Published in IN *BATALIN, I.A. (ED.) ET AL.: QUANTUM FIELD THEORY AND QUANTUM STATISTICS, VOL. 2*, 129-139. Hyperlink

Operator Ordering and the Flatness of the Universe. S.W. Hawking (Cambridge U.), Don

N. Page (Penn State U.). PRINT-80 5-0503 (PENN-STATE). Apr 1985. 21 pp. Published in Nucl.Phys. B264 (1986) 185-196, DOI: 10.1016/0550-3213(86)90478-5

The Arrow Of Time In Cosmology. S.W. Hawking (Cambridge U.). Print-80 five-0492 (CAMBRIDGE). Apr 1985.23 pp.Published in Phys.Rev. D32 (1985) 2489, DOI: 10.1103/PhysRevD.32.2489

Quantum Cosmology - Beyond Minisuperspace. J. Halliwell, S. Hawking (Cambridge U.). 1985. Published in In *Rome 1985, Proceedings, General Relativity, Pt. A*, sixty five-eighty three

The Quantum Mechanics Of The Universe. S.W. Hawking (Cambridge U.). 1985. Published in In *Geneva 1983, Proceedings, Large-scale Structure Of The Universe, Cosmology and Fundamental Physics*, 415-422

The Origin of Structure within the Universe. J.J. Halliwell, S.W. Hawking (Cambridge U. &

Munich, Max Planck Inst.). Print-80 five-0265 (CAMBRIDGE). Oct 1984. Forty eight pp. Published in Phys.Rev. D31 (1985) 1777, DOI: 10.1103/PhysRevD.31.1777

Limits On Inflationary Models Of The Universe. S.W. Hawking (Cambridge U.). Print-eighty five-0067 (CAMBRIDGE). Sep 1984. Eight pp. Published in Phys.Lett. B150 (1985) 339, DOI: 10.1016/0370-2693(80 five)90989-X

Higher Derivatives In Quantum Cosmology. 1. The Isotropic Case. S.W. Hawking, J.C. Luttrell (Cambridge U.). Print-eighty four-0711 (CAMBRIDGE). Aug 1984. Sixteen pp. Published in Nucl.Phys. B247 (1984) 250, DOI: 10.1016/0550-3213(eighty four)90380-8 Nontrivial Topologies In Quantum Gravity. S.W. Hawking (Cambridge U.). Print-84-0714 (CAMBRIDGE). Aug 1984. Sixteen pp. Published in Nucl.Phys. B244 (1984) a hundred thirty 5, DOI: 10.1016/0550-3213(eighty 4)90185-eight

Numerical Calculations Of Minisuperspace Cosmological Models. S.W. Hawking, Z.C. Wu

(Cambridge U.). Print-80 four-0913 (CAMBRIDGE). Jul 1984. 18 pp. Published in Phys.Lett. B151 (1985) 15, DOI: 10.1016/0370-2693(80 five)90815-9

The Isotropy Of The Universe. Stephen W. Hawking, Julian C. Luttrell (Cambridge U.). Print-eighty 4-0479 (CAMBRIDGE). Jun 1984. Eight pp. Published in Phys.Lett. B143 (1984) eighty three, DOI: 10.1016/0370-2693(eighty 4)90809-8

The Cosmological Constant Is Probably Zero. S.W. Hawking (Cambridge U.).Print-eighty four-0116 (CAMBRIDGE). Feb 1984. Five pp. Published in Phys.Lett. B134 (1984) 403, DOI: 10.1016/0370-2693(eighty 4)91370-4

Quantum Fluctuations As The Cause Of Inhomogeneity In The Universe. J. Halliwell, S.W. Hawking (Cambridge U.). 1984. Published in In *Moscow 1984, Proceedings, Quantum Gravity*, 509-565

The Very Early Universe. Proceedings, Nuffield Workshop, Cambridge, Uk, June 21 - July 9,

1982. G.W. Gibbons, (Ed.), S.W. Hawking, (Ed.), S.T.C. Siklos, (Ed.). 1984. Published in Cambridge, Uk: Univ. Pr. (1983) 480p

The Quantum State of the Universe. S.W. Hawking (Cambridge U.). PRINT-80 four-0117 (CAMBRIDGE). Nov 1983. 28 pp. Published in Nucl.Phys. B239 (1984) 257, DOI: 10.1016/0550-3213(eighty four)90093-2

The Unification Of Physics. S.W. Hawking (Cambridge U.). Print-80 four-0115 (CAMBRIDGE). Aug 1983. 10 pp.

Wave Function of the Universe. J.B. Hartle (Chicago U., EFI& Santa Barbara, KITP), S.W. Hawking (Cambridge U. & Santa Barbara, KITP). PRINT-83-0937 (CAMBRIDGE). Jul 983. Forty six pp. Published in Phys.Rev. D28 (1983) 2960-2975, DOI: 10.1103/PhysRevD.28.2960

Quantum Cosmology. S.W. Hawking (Cambridge U.). PRINT-80 4-0114 (CAMBRIDGE), C83-06-27.1. Jul 1983. Sixty four pp.Published in In *Les Houches 1983,

Proceedings, Relativity, Groups and Topology, Ii*, 333-379 and Preprint - HAWKING, S.W. (eighty three,REC.DEC.) 64p

Euclidean Approach To The Inflationary Universe. S.W. Hawking (Cambridge U.). Print-80 three-0318 (CAMBRIDGE). Apr 1983. 10 pp. Published in In *Cambridge 1982, Proceedings, The Very Early Universe*, 287-296 and Preprint -HAWKING, S.W. (REC.APR.Eighty 3) 12p

The Boundary Conditions For Gauged Supergravity. S.W. Hawking (Cambridge U.). Print-80 three-0317 (CAMBRIDGE). Mar 1983. Eleven pp. Published in Phys.Lett. B126 (1983) 100 seventy five, DOI: 10.1016/0370-2693(eighty 3)90585-3

Fluctuations In The Inflationary Universe. S.W. Hawking (Cambridge U.), I.G. Moss (Newcastle upon Tyne U.). PRINT-80 3-0316 (CAMBRIDGE). Dec 1982. 20 pp. Published in Nucl.Phys. B224 (1983) a hundred and eighty, DOI: 10.1016/0550-3213(eighty 3)90319-X

Thermodynamics of Black Holes in anti-De Sitter Space. S.W. Hawking (Cambridge U.), Don N. Page (Penn State U.). PRINT-80 three-0019 (CAMBRIDGE). Jul 1982. 18 pp.Published in Commun.Math.Phys. 87 (1983) 577, DOI: 10.1007/BF01208266

Positive Mass Theorems For Black Holes. G.W. Gibbons, S.W. Hawking (Cambridge U.), Gary T. Horowitz (Princeton, Inst. Advanced Study), Malcolm J. Perry (Princeton U.). Print-80 - 0505 (PRINCETON). Jul 1982. 25 pp.Published in Commun.Math.Phys. 88 (1983) 295, DOI: 10.1007/BF01213209

The Development of Irregularities in a Single Bubble Inflationary Universe. S.W. Hawking (Cambridge U.). Print-80 3-0015 (CAMBRIDGE). Jun 1982. Eight pp. Published in Phys.Lett. B115 (1982) 295, DOI: 10.1016/0370-2693(80 two)90373-2

The Unpredictability of Quantum Gravity. S.W. Hawking (Cambridge U.). Print-80 3-0017 (CAMBRIDGE). May 1982. 29 pp. Published in

Commun.Math.Phys. 87 (1982) 395-415, DOI: 10.1007/BF01206031

Bubble Collisions within the Very Early Universe. S.W. Hawking, I.G. Moss, J.M. Stewart (Cambridge U.). Print-eighty two-0180 (CAMBRIDGE). Mar 1982. 33 pp. Published in Phys.Rev. D26 (1982) 2681, DOI: 10.1103/PhysRevD.26.2681

Supercooled Phase Transitions in the Very Early Universe. S.W. Hawking, I.G. Moss (Cambridge U.). Print-eighty -0181 (CAMBRIDGE). Dec 1981. 9 pp. Published in Phys.Lett. B110 (1982) 35, DOI: 10.1016/0370-2693(80)90946-7

The Boundary Conditions Of The Universe. S.W. Hawking (Cambridge U.). PRINT-82-0179 (CAMBRIDGE). Sep 1981. Eleven pp. Published in Pontif.Acad.Sci.Scrivaria forty eight (1982) 563-574

The Cosmological Constant And The Weak Anthropic Principle. S.W. Hawking (Cambridge U.). Print-eighty two-0177 (CAMBRIDGE). Aug

1981. Nine pp. Published in In *London 1981, Proceedings, Quantum Structure Of Space and Time*, 423-432

Is The End In Sight For Theoretical Physics?. S.W. Hawking (Cambridge U.). PRINT-eighty one-0004 (CAMBRIDGE). Jan 1981. 17 pp. Published in Phys.Bull. 32 (1981) 15-17

The Loss Of Quantum Coherence Due To Virtual Black Holes. S.W. Hawking (Cambridge U.). 1981. Published in In *Moscow 1981, Proceedings, Quantum Gravity*, 19-28

Why Is The Apparent Cosmological Constant Zero? (communicate). S.W. Hawking (Cambridge U.). 1981. Published in In *Muenchen 1981, Proceedings, Unified Theories Of Elementary Particles*, 167-one hundred 75

Superspace And Supergravity. Proceedings, Nuffield Workshop, Cambridge, Uk, June sixteen - July 12, 1980.S.W. Hawking, (ed.), M. Rocek, (ed.). 1981. Published in Cambridge, Uk: Univ. Pr. (1981) 527p

Interacting Quantum Fields Around A Black Hole. S.W. Hawking (Cambridge U.). Print-eighty one-0251 (CAMBRIDGE). Dec 1980. Forty pp. Published in Commun.Math.Phys. Eighty (1981) 421, DOI: 10.1007/BF01208279

Acausal Propagation In Quantum Gravity. S.W. Hawking (Cambridge U.). PRINT-80-0866 (CAMBRIDGE), C80-04-15. Apr 1980. 22 pp. Published in In *Oxford 1980, Proceedings, Quantum Gravity 2*, 393-415

The Path Integral Approach To Quantum Gravity. S.W. Hawking (Cambridge U.). 1980. Published in In *Hawking, S.W., Israel, W.: General Relativity*, 746-789

Introductory Survey. S.W. Hawking (Cambridge U.), W. Israel (Alberta U.). 1980. Published in In *Hawking, S.W., Israel, W.: General Relativity*, 1-23

Quantum Gravitational Bubbles. S.W. Hawking, Don N. Page, C.N. Pope (Cambridge U.). Print-80-0053 (CAMBRIDGE). Oct 1979. 33

pp. Published in Nucl.Phys. B170 (1980) 283-306, DOI: 10.1016/0550-3213(80)90151-0

Yang-turbines Instantons And The S Matrix. S.W. Hawking, C.N. Pope (Cambridge U.). Print-79-0654 (CAMBRIDGE). Apr 1979. 32 pp. Published in Nucl.Phys. B161 (1979) 90 three, DOI: 10.1016/0550-3213(seventy nine)90128-7

Space-Time Foam. S.W. Hawking (Cambridge U.). Print-seventy nine-0038 (CAMBRIDGE). Jan 1979. 24 pp. Published in Nucl.Phys. B144 (1978) 349-362, DOI: 10.1016/0550-3213(78)90375-9

Gravitational Multi – Instantons. G.W. Gibbons, S.W. Hawking (Cambridge U.). Print-seventy nine-0042 (CAMBRIDGE). Jan 1979. 6 pp. Published in Phys.Lett. B78 (1978) 430, DOI: 10.1016/0370-2693(78)90478-1

Symmetry Breaking By Instantons In Supergravity. S.W. Hawking, C.N. Pope (Cambridge U.). Print-seventy nine-0043 (CAMBRIDGE). Jan 1979. 22 pp. Published in

Nucl.Phys. B146 (1978) 381, DOI: 10.1016/0550-3213(seventy eight)90073-1

The Propagation Of Particles In Space-time Foam. S.W. Hawking, Don N. Page, C.N. Pope (Cambridge U.). 1979. Published in Phys.Lett. B86 (1979) one hundred seventy five-178, DOI: 10.1016/0370-2693(seventy nine)90812-eight

Classification of Gravitational Instanton Symmetries. G.W. Gibbons, S.W. Hawking (Cambridge U.). 1979. Published in Commun.Math.Phys. 66 (1979) 291-310, DOI: 10.1007/BF01197189

Relativity Today. S. Hawking (Cambridge U.), W. Israel (Alberta U.). 1979. Published in New Sci. 81 (1979) 761-763

General Relativity. An Einstein Centenary Survey Pt 1. S.W. Hawking (Cambridge U.), W. Israel (Alberta U.). 1979. Published in Cambridge, United Kingdom: Univ.Pr.(1979) 919p

Theoretical Advances In General Relativity. S.W. Hawking (Cambridge U.). Print-79-0595 (CAMBRIDGE). Nov 1978. Sixteen pp.

Euclidean Quantum Gravity. Stephen W. Hawking (Cambridge U.). PRINT-seventy eight-0745 (CAMBRIDGE), C78-07-10.1-2. Jul 1978. 30 pp. Published in NATO Adv.Study Inst.Ser.B Phys. Forty 4 (1979) 145

Path Integrals and the Indefiniteness of the Gravitational Action. G.W. Gibbons (Munich, Max Planck Inst. & Cambridge U.), S.W. Hawking, M.J. Perry (Cambridge U.). PRINT-78-0375 (CAMBRIDGE). Apr 1978. 14 pp. Published in Nucl.Phys. B138 (1978) 141, DOI: 10.1016/0550-3213(78)90161-X

Quantum Gravity and Path Integrals. S.W. Hawking (Cambridge U. & Caltech). 1978. Published in Phys.Rev. D18 (1978) 1747-1753, DOI: 10.1103/PhysRevD.18.1747

Generalized Spin Structures in Quantum Gravity. S.W. Hawking, C.N. Pope (Cambridge U.). Print-78-0374 (CAMBRIDGE). Nov 1977. 6

pp. Published in Phys.Lett. B73 (1978) 42-forty four, DOI: 10.1016/0370-2693(seventy eight)90167-3

Cosmological Event Horizons, Thermodynamics, and Particle Creation. G.W. Gibbons, S.W. Hawking (Cambridge U.). 1977. Published in Phys.Rev. D15 (1977) 2738-2751, DOI: 10.1103/PhysRevD.15.2738

The Quantum Mechanics of Black Holes. S.W. Hawking. 1977. Published in Sci.Am. 236 (1977) 34-40 9, DOI: 10.1038/scientificamerican0177-34

Black Holes and Unpredictability. S.W. Hawking (Cambridge U.). PRINT-seventy seven-0292 (CAMBRIDGE). Dec 1976. 6 pp.Published in Phys.Bull. 29 (1978) 23-24

Gravitational Instantons. S.W. Hawking (Cambridge U.). Print-77-0294 (CAMBRIDGE). Dec 1976. Eight pp. Published in Phys.Lett. A60 (1977) eighty one, DOI: 10.1016/0375-9601(seventy seven)90386-three

Zeta Function Regularization of Path Integrals in Curved Space-Time. S.W. Hawking (Cambridge U.). PRINT-77-0293 (CAMBRIDGE). Dec 1976. 29 pp. Published in Commun.Math.Phys. Fifty 5 (1977) 133, DOI: 10.1007/BF01626516

Action Integrals and Partition Functions in Quantum Gravity. G.W. Gibbons, S.W. Hawking (Cambridge U.). PRINT-76-0995 (CAMBRIDGE). Sep 1976. 14 pp. Published in Phys.Rev. D15 (1977) 2752-2756, DOI: 10.1103/PhysRevD.15.2752

Gamma rays from primordial black holes. Don N. Page, S.W. Hawking. May 1976. 7 pp. Published in Astrophys.J. 206 (1976) 1-7, DOI: 10.1086/154350

Breakdown of Predictability in Gravitational Collapse. S.W. Hawking (Cambridge U. & Caltech). 1976. Published in Phys.Rev. D14 (1976) 2460-2473, DOI: 10.1103/PhysRevD.14.2460

Path Integral Derivation of Black Hole Radiance. J.B. Hartle, S.W. Hawking (UC, Santa Barbara & Caltech & Cambridge U.). 1976. Published in Phys.Rev. D13 (1976) 2188-2203, DOI: 10.1103/PhysRevD.13.2188

Black Holes and Thermodynamics. S.W. Hawking (Caltech & Cambridge U.). 1976. Published in Phys.Rev. D13 (1976) 191-197, DOI: 10.1103/PhysRevD.Thirteen.191

A New Topology for Curved Space-Time Which Incorporates the Causal, Differential, and Conformal Structures. S.W. Hawking (Cambridge U. & Caltech), A.R. King, P.J. Mccarthy. 1976. Published in J.Math.Phys. 17 (1976) 174-181, DOI: 10.1063/1.522874

Chapter 11: Using Manner Of Black Holes

S.W. Hawking (Cambridge U.). Aug 1975. 22 pp. Published in Commun.Math.Phys. Forty three (1975) 199-220, Erratum-ibid. Forty six (1976) 206-206, DOI: 10.1007/BF02345020

Black hollow explosions. S.W. Hawking (Cambridge U.). Mar 1974. 2 pp. Published in Nature 248 (1974) 30-31, DOI: 10.1038/248030a0

Black holes in the early Universe. Bernard J. Carr, S.W. Hawking (Cambridge U., Inst. Of Astron. & Cambridge U., DAMTP). Feb 1974. 17 pp. Published in Mon.Not.Roy.Astron.Soc. 168 (1974) 399-415. DOI: 10.1093/mnras/168.2.399

Causally non-save you space-times. S.W. Hawking, R.K. Sachs. 1974. Published in Commun.Math.Phys. 35 (1974) 287-296, DOI: 10.1007/BF01646350

A Variational principle for black holes. S.W. Hawking. 1973. Published in

Commun.Math.Phys. 33 (1973) 323-334, DOI: 10.1007/BF01646744

The Four felony tips of black hollow mechanics. James M. Bardeen (Yale U.), B. Carter, S.W. Hawking (Cambridge U.). 1973. Published in Commun.Math.Phys. 31 (1973) 161-one hundred and seventy, DOI: 10.1007/BF01645742

The Large scale structure of area-time. S.W. Hawking, G.F.R. Ellis. 1973. 391 pp. Published in Cambridge University Press, Cambridge, 1973. ISBN-13: 978-0521099066

The rotation and distortion of the universe. C.B. Collins, S.W. Hawking. Jan 1973, Published in Mon.Not.Roy.Astron.Soc. 162 (1973) 307-320

Why is the Universe isotropic?. C.B. Collins, S.W. Hawking (Cambridge U., DAMTP & Cambridge U.). Sep 1972. 18 pp. Published in Astrophys.J. A hundred and eighty (1973) 317-334, DOI: 10.1086/151965

Solutions of the Einstein-Maxwell equations with many black holes. J.B. Hartle, S.W. Hawking. Jun 1972. Published in Commun.Math.Phys. 26 (1972) 87-one hundred and one, DOI: 10.1007/BF01645696

Energy and angular momentum circulate a black hollow. S.W. Hawking (Cambridge U., DAMTP), J.B. Hartle (UC, Santa Barbara). 1972. Published in Commun.Math.Phys. 27 (1972) 283-290, DOI: 10.1007/BF01645515

Black holes in the Brans-Dicke principle of gravitation. S.W. Hawking (Cambridge U.). 1972. Published in Commun.Math.Phys. 25 (1972) 167-171, DOI: 10.1007/BF01877518

Theory of the detection of quick bursts of gravitational radiation. G.W. Gibbons, S.W. Hawking (Cambridge U., DAMTP). 1972. Published in Phys.Rev. D4 (1971) 2191-2197, DOI: 10.1103/PhysRevD.Four.2191

Black holes in famous relativity. S.W. Hawking (Cambridge U.). Oct 1971. Published in

Commun.Math.Phys. 25 (1972) 152-166, DOI: 10.1007/BF01877517

Gravitational radiation from colliding black holes. S.W. Hawking (Cambridge U.). Mar 1971. Published in Phys.Rev.Lett. 26 (1971) 1344-1346, DOI: 10.1103/PhysRevLett.26.1344

Evidence for black holes in binary superstar structures. S.W. Hawking, G.W. Gibbons. 1971. Published in Nature 232 (1971) 465, DOI: 10.1038/232465a0

The Definition and prevalence of singularities in famous relativity. Stephen Hawking. 1971. Published in Lect.Notes Math. 209 (1971) 275-279

Stable and traditional houses in favored relativity. Stephen Hawking (Cambridge U., Inst. Of Astron.). 1971. Published in Gen.Rel.Grav. 1 (1971) 393-4 hundred, DOI: 10.1007/BF00759218

Gravitationally collapsed devices of very low mass. Stephen Hawking. 1971. Published in

Mon.Not.Roy.Astron.Soc. 152 (1971) 75. DOI: 10.1093/mnras/152.1.Seventy five

The Singularities of gravitational collapse and cosmology. S.W. Hawking (Cambridge U.), R. Penrose (Birkbeck Coll.). Jan 1970. 20 pp. Published in Proc.Roy.Soc.Lond. A314 (1970) 529-548, DOI: 10.1098/rspa.1970.0021

The conservation of rely in present day relativity. S. Hawking (Cambridge U., DAMTP). 1970. Published in Commun.Math.Phys. 18 (1970) 301-306, DOI: 10.1007/BF01649448

Singularities in collapsing stars and universes. Stephen Hawking, Dennis Sciama. 1969. Published in Comments Astrophys. Space Phys. 1 (1969) 1

On the Rotation of the universe. S.W. Hawking (Cambridge U., Inst. Of Astron.). Sep 1968. Thirteen pp. Published in Mon.Not.Roy.Astron.Soc. 142 (1969) 129-141.

Gravitational radiation in an expanding universe. Stephen Hawking (Cambridge U.,

DAMTP). Apr 1968. Published in J.Math.Phys. Nine (1968) 598-604, DOI: 10.1063/1.1664615

The Cosmic black body radiation and the existence of singularities in our universe. G.F.R. Ellis, Stephen Hawking. 1968. Published in Astrophys.J. 152 (1968) 25, DOI: 10.1086/149520

The Existence of cosmic time functions. Stephen Hawking (Cambridge U., DAMTP). 1968. Published in Proc.Roy.Soc.Lond. A308 (1968) 433-435. DOI: 10.1098/rspa.1969.0018

The occurrence of singularities in cosmology. III. Causality and singularities. Stephen Hawking (Cambridge U., DAMTP). 1967. Published in Proc.Roy.Soc.Lond. A300 (1967) 187-201, DOI: 10.1098/rspa.1967.0164

Perturbations of an growing universe. S.W. Hawking (Cambridge U., DAMTP). Feb 1966. Eleven pp. Published in Astrophys.J. One hundred 45 (1966) 544-554, DOI: 10.1086/148793

Singularities in the universe. S.W. Hawking. 1966. Published in Phys.Rev.Lett. 17 (1966) 444-445, DOI: 10.1103/PhysRevLett.17.444

Helium production in anisotropic massive bang universes. Stephen Hawking, J.R. Tayler (Cambridge U., DAMTP). 1966. Published in Nature 209 (1966) 1278-1279, DOI: 10.1038/2091278a0

The Occurrence of singularities in cosmology. Stephen Hawking (Cambridge U., DAMTP). 1966. Published in Proc.Roy.Soc.Lond. A294 (1966) 511-521, DOI: 10.1098/rspa.1966.0221

The Occurrence of singularities in cosmology. II. Stephen Hawking (Cambridge U., DAMTP). 1966. Published in Proc.Roy.Soc.Lond. A295 (1966) 490-493, DOI: 10.1098/rspa.1966.0255

Singularities and the geometry of area-time. Stephen Hawking. 1966. DOI: 10.1140/epjh/e2014-50013-6

Singularities in homogeneous international models. Stephen Hawking, G.F.R. Ellis (Cambridge U., DAMTP & Cambridge U.). Jun

1965. Published in Phys.Lett. 17 (1965) 246-247, DOI: 10.1016/0031-9163(sixty five)90510-X

On the Hoyle-Narlikar principle of gravitation. Stephen Hawking (Cambridge U., DAMTP). Feb 1965. Published in Proc.Roy.Soc.Lond. A286 (1965) 313-319, DOI: 10.1098/rspa.1965.0146

Occurrence of singularities in open universes. Stephen Hawking (Cambridge U., DAMTP). 1965. Published in Phys.Rev. Ett. 15 (1965) 689-690, DOI: 10.1103/PhysRevLett.15.689

Properties of Expanding Universes. Stephen Hawking, PhD Thesis. 1965, DOI: 10.17863/CAM.11283

Chapter 12: George And The Big Bang

Stephen and Lucy Hawking's fourth book about the adventures of George and Annie.

Just as thrilling a ebook as all of the ones earlier than it, George and the Unbreakable Code moreover delves deeper and covers subjects and thoughts that the meant readers won't commonly maintain in thoughts.

Overall, a terrific ebook as constantly brought from the proper Hawking duo.

The zero.33 ebook within the collection, with a entire color set of pics, illustrating the marvel of the cosmos, George and the Big Bang, follows George's adventures in the universe.

George's Cosmic Treasure Hunt

The 2d inside the collection of children's books written with the aid of way of Stephen and Lucy Hawking. George and Annie, the middle-university cosmologists, move lower back in this sequel to George's Secret Key to the Universe. You can order your replica at amazon.Com or amazon.Co.United country.

George's Secret Key to the Universe

The first in a sequence of kid's books that melds cosmology and journey, co-authored through Professor Hawking and his daughter Lucy.

George's exquisite buddy Annie goals help. Her scientist father, Eric, is strolling on a place assignment - and it's miles all going wrong. A robotic has landed on Mars, however is behaving very oddly. And now Annie has located a few aspect bizarre on her dad's incredible-computer.

The Universe in a Nutshell

In two hundred pretty illustrated pages, Hawking is pushing the frontiers of famous physics past relativity and quantum concept, beyond superstring concept and imaginary time, right into a dizzying new worldwide of M-concept and branes. Black Holes and Baby Universes

In his first collection of essays and other portions - on subjects that variety from warmly personal to the wholly clinical- Stephen Hawking is observed out variously because of the reality the scientist, the character, the involved global citizen, and - as typically - the rigorous and inventive philosopher.

On the Shoulders of Giants

On the Shoulders of Giants tells a compelling tale, using real papers from Einstein, Copernicus, Galilei, Kepler and Newton. Theoretical physicist Stephen Hawking explains how those works modified the path of technology, ushering astronomy and physics out of the Middle Ages and into the current-day worldwide.

The Large Scale Structure of Space-time

A textbook for Physicists, this 1973 ebook explores predictions of Einstein's General Theory of Relativity: first, that the last destiny of many large stars is to undergo gravitational disintegrate and to vanish from view, leaving inside the again of a 'black hole' in location; and secondly, that there will exist singularities in area-time itself.

God created the Integers

God Created The Integersis Stephen Hawking's private preference of the first-class mathematical works in statistics. The book consists of landmark discoveries spanning 2500 years and representing the art work of mathematicians collectively with Euclid, Georg Cantor, Kurt Godel, Augustin Cauchy, Bernard Riemann and Alan Turing.